Climate Policy Uncertainty and Investment Risk

INTERNATIONAL ENERGY AGENCY

The International Energy Agency (IEA) is an autonomous body which was established in November 1974 within the framework of the Organisation for Economic Co-operation and Development (OECD) to implement an international energy programme.

It carries out a comprehensive programme of energy co-operation among twenty-six of the OECD thirty member countries. The basic aims of the IEA are:

- To maintain and improve systems for coping with oil supply disruptions.
- To promote rational energy policies in a global context through co-operative relations with non-member countries, industry and international organisations.
- To operate a permanent information system on the international oil market.
- To improve the world's energy supply and demand structure by developing alternative energy sources and increasing the efficiency of energy use.
- To assist in the integration of environmental and energy policies.

The IEA member countries are: Australia, Austria, Belgium, Canada, the Czech Republic, Denmark, Finland, France, Germany, Greece, Hungary, Ireland, Italy, Japan, the Republic of Korea, Luxembourg, the Netherlands, New Zealand, Norway, Portugal, Spain, Sweden, Switzerland, Turkey, the United Kingdom and the United States. The European Commission takes part in the work of the IEA.

ORGANISATION FOR ECONOMIC CO-OPERATION AND DEVELOPMENT

The OECD is a unique forum where the governments of thirty democracies work together to address the economic, social and environmental challenges of globalisation. The OECD is also at the forefront of efforts to understand and to help governments respond to new developments and concerns, such as corporate governance, the information economy and the challenges of an ageing population. The Organisation provides a setting where governments can compare policy experiences, seek answers to common problems, identify good practice and work to co-ordinate domestic and international policies.

The OECD member countries are: Australia, Austria, Belgium, Canada, the Czech Republic, Denmark, Finland, France, Germany, Greece, Hungary, Iceland, Ireland, Italy, Japan, Korea, Luxembourg, Mexico, the Netherlands, New Zealand, Norway, Poland, Portugal, the Slovak Republic, Spain, Sweden, Switzerland, Turkey, the United Kingdom and the United States. The European Commission takes part in the work of the OECD.

Foreword

In the coming years, the world will need more electric power, but fewer greenhouse gas emissions. How do we reconcile these seemingly contradictory objectives? Providing for abundant, affordable, clean energy will require considerable investment in new power generation - more than USD 11 trillion to 2030 in the IEA *World Energy Outlook 2006* Reference Scenario. This investment will need to be timely, respond to market signals, promote the development and deployment of affordable new technologies and have a reduced carbon footprint. Yet there are many other places where investors can put their capital. How can we be sure that adequate investment decisions will be made?

Investment in the power sector depends on various factors – the national economy, government policies and regulations, the energy market, technological advances, as well as operations and maintenance of new technologies – which all can add risk to the portfolio of power investors. Of these various factors, uncertainty has become an increasing concern of investors in power plants. In particular, energy markets and government climate change policies are increasingly unpredictable, making the cash inflow of an investment project less certain.

This book performs an extensive quantitative evaluation of the impacts of energy market uncertainty and climate change policy uncertainty. The results lead to recommendations that are intended to assist government policy makers in designing better climate change policies that motivate energy market players to make timely investments in appropriate technologies with more certainty. At stake are energy security, economic growth and environment protection.

Claude Mandil
Executive Director

Acknowledgements

Climate Policy Uncertainty and Investment Risk was researched and written by Dr. William Blyth, a consultant of the IEA; and Dr. Ming Yang and Dr. Richard Bradley of the Energy Efficiency and Environment Division of the IEA.

The authors would like to acknowledge the support of colleagues at the IEA including Richard Baron, Cedric Philibert, Jonathan Coony, and Maria Argiri. Charlie Clark, Tom Wilson and Adam Diamant at EPRI provided inspiration for the project and valuable input to the modelling methodology. Derek Bunn at London Business School has provided intellectual support to the work. Andy Read of E.ON UK, Nikhil Venkateswaren of RWE npower and Eliano Russo and Giuseppe Montesano of Enel have all provided very useful input and feedback on the results, as have many others from companies listed in Appendix 3. Bev Darkin and Kirsty Hamilton provided ideas during the project, and Chatham House provided administrative and research support and hosted the associated workshops. Acknowledgement is also due to Ms. Wanda Ollis for her professional proof-reading of the manuscript of the book.

Unless otherwise noted, the figures and tables in this report are generated by the authors during the study.

Questions and comments should be addressed to:
Dr. Ming Yang
Energy and environment economist
Energy efficiency and environment division
International Energy Agency
Email: ming.yang@iea.org

Table of contents

List of tables

List of figures

Executive Summary

Introduction

Business routinely deals with risk and uncertainty in decision making and will continue to do so in the face of climate change policy uncertainty. Risk is not inherently a bad thing. It is by taking calculated risks that companies aim to make profits in excess of their cost of capital. Nevertheless, sustained additional risk raises the cost of capital, and will alter investment decisions.

The response of business to policy risk is important for the effectiveness of both climate and energy policy. This analysis suggests that in some cases business decisions will be different under conditions of policy uncertainty, and that therefore policies may need to be designed differently or made more stringent than expected. This book can assist policy makers to understand the nature of the investment decision and how it is influenced by policy uncertainty. It can therefore contribute to more effective policy design.

This book provides insights into how investment behaviour in the power sector may be affected by climate change policy uncertainty. It looks at investment in coal, gas, oil, nuclear and carbon capture and storage technologies.

Key Message 1

The analysis suggests that it is unlikely in most markets that climate policy uncertainty would pose a serious threat to overall generation capacity levels in the long run.

This is because, if climate policy is set over sufficiently long timescales, the total risk will generally be dominated by fuel price risk, with climate policy risk contributing relatively little to the total risk profile of the investments.[1] Fuel price risks may however be less significant in some markets (*e.g.* Australia) where gas prices are perceived to be more stable than assumed in the model used here.

1. The exceptions to this are: (1) nuclear investments that require a significant carbon price to make them financially viable where coal plant is the marginal plant in the merit order; and (2) where CO_2 prices significantly increase the price of coal.

On the other hand, climate policy uncertainty does weaken investment incentives for low-carbon technologies. Uncertainty could also lead to investment choices that would appear sub-optimal in a world of greater policy certainty. Unfavourable effects of policy uncertainty could include, extending the life of existing plant rather than investing in more efficient new plant, modest increases in electricity prices, and the creation of investment cycles that may exacerbate short-term peaks and troughs in generation capacity.

It is certain that in the long run, we will have to find ways of satisfying our energy needs with near-zero net emissions of greenhouse gases in order to avoid the worst damage from climate change. This will require an almost complete turnover in the world's energy infrastructure.

What is uncertain, however, is when this transition will start in earnest, and how quickly it will proceed. The rate of transition to a near-zero-emitting energy infrastructure will determine the total stock of greenhouse gases emitted to the atmosphere and the degree of climate change to which the planet will be committed. The rate of transition will be constrained by the costs of transition, vested interests in the *status quo* and the level of political and popular will to drive change.

There have been over 15 years of international climate change negotiations, and many important climate change policy initiatives have been undertaken in many countries. Nevertheless, compared to the scale of the task, climate change policy and programmes are still in their infancy. During these early stages, policy uncertainty is high, perhaps at its peak. As action is taken to reduce emissions of greenhouse gases, this should strengthen the credibility and effectiveness of climate change policy, which in turn will improve the business case for further action to cut emissions. Until this self-reinforcing pattern is established, investment decisions will need to be taken under considerable uncertainty.

Investment is driven by expectations of future returns, which will depend on future market conditions. Changes in current market conditions will affect future expectations, but will not entirely determine them. For example, an estimate of the future price of a commodity will take into account the current spot price, but will also incorporate many other factors relating to expected future developments. This also applies for a newly introduced climate change policy. Since there is a possibility that the policy could change, future expectations are not simply determined by the current status of the policy.

Approach to quantifying uncertainty

The analysis in this book provides a quantification of the investment risk created by policy uncertainty. The approach puts climate policy uncertainty on the same footing as other investment risks faced by power companies and enables policy makers to determine its relative importance. It also provides a useful conceptual framework, moving away from a discussion of investment "barriers" towards an understanding of the likely risk management responses that companies may adopt in the face of climate policy uncertainty.

The principle by which policy uncertainty can be translated into investment risk is straightforward. Policy uncertainty creates an uncertain outcome in the cash flow of a project in which the company is proposing to invest. Faced with this uncertain cash flow, the company may have the option to wait until the policy uncertainty is resolved. On the one hand, by waiting, it may be able to avoid the worst financial outcomes by tailoring its investment decisions in response to policy developments. In the meantime, waiting rather than investing immediately could result in foregoing income. The value of waiting therefore has to be balanced against the opportunity cost of waiting.

The level of risks will depend on the type and design of the climate change policy being considered. We characterise all climate policies in terms of an effective carbon price, so that policy uncertainty is translated into a carbon price uncertainty. This approach is most directly applicable to taxes and trading schemes in which an explicit price is established, but could also be applied to any policy for which the price of carbon is a proxy for the costs of compliance with the policy.

We model two elements of carbon price uncertainty, a one-off jump in price to represent a possible change in policy at some time in the future, and an annual fluctuation to represent price volatility. In general, we find that policy uncertainty dominates the risk premium, whereas price volatility plays quite a small role. Whilst carbon taxes would eliminate price volatility, they are still prone to unexpected changes in levels, so do not necessarily perform much better in terms of reducing uncertainty than an emissions trading scheme. The techniques used in this analysis could be extended to look at the case where there is currently no effective price of carbon. In this case, a key uncertainty would be the timing of the introduction of climate change policy. This case is discussed qualitatively.

The analysis does not aim to include a full range of investment risks faced by companies. The main emphasis is on carbon dioxide (CO_2) price risks, but fuel price risks are also included to give a comparison and some context to the analysis.

Risk premiums

Risk, whether associated with fuel price uncertainty or carbon price uncertainty requires net returns to be higher than would be necessary where there was no uncertainty. This higher net return is called the "risk premium" for either fuel price uncertainty or carbon price uncertainty. The risk premiums depend on the technology being considered, the market context in which the company operates and the details of the climate change policy mechanism being considered.

The quantitative analysis in this book looks at the risk premium associated with uncertainty for an existing or a proposed new policy. These risk premiums are derived from a consideration of the flexibility that companies have to defer investment and wait for additional information that could improve the outcome of their investment decision.

For all generation technologies the climate change policy risk premium depends on how long there is left for the policy to run. The fewer the number of years remaining until an expected change in policy, the greater the risk premium associated with policy uncertainty. This assumes that there is no visibility at all about future climate policy before the end of the existing policy. Figure 1 shows the risk premiums in terms of additional capital investment costs (USD/kW) that are associated with uncertainties of energy price and carbon price.

Key Message 2

The results indicate that climate policy risks may be brought down to modest levels compared to other risks if policy is set over a sufficiently long timescale into the future.

One method for reducing the effects of uncertainty is to try to shift the expected policy change further out into the future (*i.e.* to set policy over longer timescales). Investment risk premiums are significantly lower when the price jump representing the policy uncertainty is shifted from five years in the future out to 10 years in the future. The period of 5-15 years into the future is the key period

FIGURE 1

Range of risk premiums for new investments created by uncertainty*

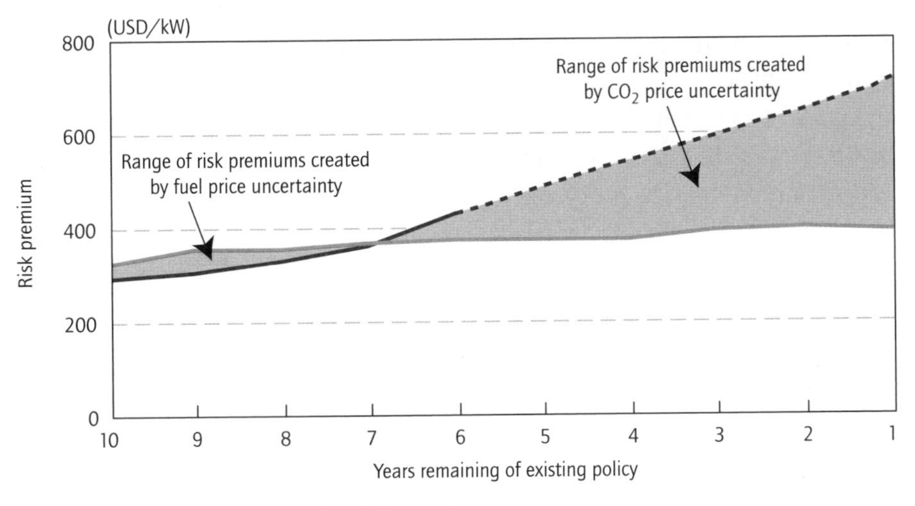

*The dotted line is an extrapolation of modelling results.

over which a planned new power generation needs to recoup the majority of its investment. Creating visibility of prices and policy design decisions over this timeframe has been a consistent theme in all the discussions with companies during the analysis for this book. Ideally, this should be done as a continuous process, so that price visibility is always maintained at least 10 years ahead. This would also help to avoid creating cyclical investment incentives. Setting aspirational targets for the very long term (*e.g.* to 2050) without providing milestones for this key mid-term period does not significantly help reduce investment risk premiums.

However, simply defining a policy goal over a 10-year period is not sufficient to overcome policy risk. There also needs to be credibility that the policy will not be changed during this period. Policy credibility is not entirely within the control of individual national governments. If similar actions are not committed to by other countries, then there will always be a risk of backtracking on the grounds of maintaining competitiveness. Credibility relies on the accumulation of the experiences and actions taken by governments and companies internationally.

Key Message 3

Climate policy risks will be important for investments for which climate policy is a dominant economic driver.

The example taken in this book is carbon capture and storage (CCS). This technology separates CO_2 in the combustion process, transports it and pumps it into geological storage sites underground. The process requires energy itself, and therefore reduces the efficiency of power generation. The only rationale for the technology is as a way of achieving compliance with climate change policy. Uncertainty in climate policy therefore strongly affects the economic case for investment. The technology cost assumptions used in the model mean that the price of carbon would need to be USD 38/tCO_2 in order to stimulate investment in CCS under conditions of price certainty (*i.e.* using a normal discounted cash flow analysis). To stimulate investment in the technology under conditions of policy uncertainty, the effective price of carbon would need to rise to between USD 44/tCO_2 (when policy is set 10 years ahead) and USD 52/tCO_2 (when policy is set 5 years ahead). This shows that the greater the level of policy uncertainty, the less effective climate change policies will be at incentivising investment in low-emitting technologies.

Carbon capture and storage acts as a good hedge for coal generation against future increases in CO_2 price. The ability to retrofit CCS to an existing plant in response to policy changes reduces the risks for coal generation, and may accelerate investment in a new coal plant. In this study, we have not considered the technical risks of CCS. However, we conclude that as the technical risks are addressed, the option value of CCS will be further enhanced, further accelerating investment in coal.

Companies operating in a price-regulated market will have a different risk exposure from those indicated above. The key source of climate policy risk in these markets is whether the regulator will allow compliance costs to be passed through to the consumers. In coal-dominated markets in particular, operating costs are low and carbon emissions are high. Price increases associated with the introduction of stringent climate policies could therefore be strong, and it is not necessarily certain that regulators will allow everything to be passed through.

Key Message 4

The closer in time a company is to a change in policy, the greater the policy risk will be, and the greater the impact on investment decisions. If there are only a few years left before a change, policy uncertainty could become a dominant risk factor. This may occur as we approach the end of the current commitment period and investors face considerable uncertainty about the structure, stringency and timing of a post-2012 mitigation regime.

It is therefore possible that there could be a period with very little new investment in the lead-up to the start of a new policy (or a new phase of an existing policy) if key parameters such as tax rates or emissions caps are not announced well in advance. This could create problems, particularly if it exacerbates other factors in the market such as plant closures due to environmental regulation of nitrogen oxide (NO_x) and sulphur dioxide (SO_2) or nuclear phase-out for example.

Key Message 5

Risk premiums could be reduced if price constraints could be established that limited future price variability, either for a tax or a trading scheme. These constraints would have to be credible over a long period, with a very low probability that prices would move outside these constraints.

Price caps on their own, in the absence of a corresponding price floor, create an asymmetrical price risk. This would marginally improve the investment case for a high-emitting coal plant and making the investment case for low-emitting technologies marginally worse. It is possible however, that with prices capped, the political will to set more aggressive climate change targets would be increased, restoring the case for investment in low carbon technologies. Conversely, price floors on their own would improve the investment case for low carbon technologies and make the investment case for a high-emitting plant worse.

Key Message 6

Companies will generally be confident in committing capital to projects, even in an uncertain environment, as long as they can establish a competitive advantage over other players in the market. When it comes to regulatory risk, this requires that policy makers establish clear rules, and that companies can be confident that these rules will be applied consistently to all market players, irrespective of ownership structure.

Climate change policies affect companies in a number of different ways beyond just uncertainty about the carbon price. For example, in an emissions trading scheme, the allocation of free allowances to incumbents based on their historical emissions can have important financial implications for those companies. Transparency on these types of policy details, and the criteria and processes by which the rules may get reviewed and implemented can also be very important in helping to manage policy risk.

Climate policy risks are starting to be recognised by financial markets. Initiatives such as the Carbon Disclosure Project represent major groupings of institutional investors with assets of over USD 30 trillion. These groups have started to press for clear statements from big publicly owned companies of their exposure to risk from climate change and climate change policies. This type of activity raises the pressure on companies to consider policy risk in a more open and structured way. The risks are expected to increase as the pace of change increases.

Use and interpretation of results

This book demonstrates the importance of incorporating risk into policy analysis in order to understand investment behaviour. Given the broad geographical scope of the work (aiming to be relevant to all IEA countries), the purpose of the quantitative analysis is to demonstrate a conceptual framework for thinking about investment under uncertainty, and to give an illustration of the scale of the effects of policy uncertainty. It is not intended to provide a complete representation of investment risk - the quantitative results depend critically on the input assumptions, which are necessarily methodological choices. For detailed policy analysis at the national level, the assumptions would need to be more closely tailored to particular national circumstances. Other drivers of investment decisions, such as compliance with other (non-climate related) environmental regulations can be important, and should be incorporated into the analysis. Ideally, the results from this type of stochastic optimisation analysis should be combined with macroeconomic models in order to capture broader dynamic impacts.

INTRODUCTION

Getting the right type of investment in infrastructure for energy supply and consumption is a minimum requirement to enable the transition towards a sustainable energy system. One of the key tasks of climate change policy makers is therefore to create incentives to encourage the necessary investments to be undertaken. However, the translation of climate policies into clear investment signals is not straightforward. Energy infrastructure investments occur in a highly dynamic context, where climate policy is one of many different risk factors to take into account.

Policy uncertainty is an important example of how stated policy aims may not translate easily into investment action. Uncertainty has consistently been raised by business in discussion with governments and regulators as a cause for concern and a potential barrier to investment, as described in the quotes below (see also for example, Hamilton, 2005; Environmental Finance, 2005; Wolf, 2004):

Significant uncertainties that are unclear or unmanageable lead us to make decisions not to invest in projects affected by such uncertainties. One uncertainty that fits this description is the risk of adverse governmental laws or actions. In general, we choose to invest in markets where the regulator has made the commitment to develop rules that are transparent, stable and fair. The rules do not have to be exactly what we want, so long as we can operate within their framework. Consequently, we look for markets where the rules of competition are clear, encouraged and relatively stable.

Source: Geoffrey Roberts, President & CEO, Entergy Wholesale Operations, U.S. Senate Hearing on S.764, June 19, 2001[2].

Probably the greatest uncertainty for investors in new power plants will be controls on future carbon dioxide emission. The unknown value of carbon emissions permits and the mechanism chosen to allocate permits will become a very large and potentially critical uncertainty in power generation investment. This uncertainty will grow in the future, as restrictions on levels of carbon dioxide emissions beyond the first commitment period of the Kyoto Protocol are unknown.

Source: IEA, 2003b p31 Power Generation Investment in Electricity Markets.

2. http://energy.senate.gov/hearings/testimony.cfm?id=548&wit_id=136.

In addition, viewed from the perspective of the climate policy makers, the signals are already clear: there is a need to move towards zero-emitting technologies as quickly as is economically and socially feasible. The uncertainties about how rapidly this can be done, and how the burden of action will be divided between different groups is undoubtedly recognised. However, there is still a lack of quantitative analysis that allows communication between investors who face the financial risks associated with uncertain climate change policy and policy makers who are trying to decide how quickly they can push change.

This book begins to fill the gaps. It combines quantitative modelling analysis with qualitative analysis reflecting discussions held among the various stakeholders. We explore how climate policy uncertainty may affect patterns of investment in the power sector and identify opportunities to reduce adverse effects through appropriate policy design.

The scope of the analysis examines large-scale centralised power generation options, including coal, gas, oil and nuclear power generation. We do not look at renewable energy options or investments in other parts of the energy system, such as: transmission, distribution, distributed generation or downstream energy efficiency. These other aspects may be covered in future work. Given the broad geographical scope of the work (aiming to be relevant to all IEA countries), the purpose of the quantitative analysis is to demonstrate a conceptual framework for thinking about investment under uncertainty, and to give an illustration of the scale of the effects of policy uncertainty. It is not intended to provide a complete representation of investment risk as the quantitative results depend on the input assumptions, which are necessarily stylised. The quantitative analysis provides useful input to illuminate discussions on policy design.

Chapter 1 contextualises the analysis by providing an overview of the origins of climate change policy uncertainty and by reviewing some of the theoretical approaches to assessing and managing risk. Chapter 2 goes on to introduce the analytical framework used to quantify the effects of uncertainty on investment decision making, while Chapter 3 presents the quantitative results of this analysis. Chapter 4 then broadens the discussion to incorporate the views and experiences of real companies operating in different types of markets in different countries. Chapter 5 brings together key messages for policy makers.

Climate change uncertainties

Fundamentally, the uncertainties in the field of climate change arise from uncertainties about the potential physical impacts of an increase in greenhouse gas concentrations in the atmosphere, together with uncertainties about the cost of reducing emissions of greenhouse gases to slow down this accumulation in the atmosphere. The two most uncertain properties that control the climate system's response over several decades to increases in greenhouse gas concentrations are climate sensitivity (the increase in global mean temperature in response to a doubling of atmospheric concentrations) and the rate of heat uptake by the deep ocean (Forest *et al.*, 2002). The climate sensitivity parameter is typically quoted as being in the range 1.5-4.5°C, the range quoted in the Intergovernmental Panel on Climate Change (IPCC) Third Assessment Report (IPCC, 2001). The rate of heat uptake by the ocean remains poorly specified and affects the potential range of values for climate sensitivity. This range for mean temperature response has a strong effect on the emissions reductions required to achieve policies framed in terms of meeting maximum warming rates (Caldeira *et al.*, 2003).

The uncertainty in the above figure for temperature response feeds into an even wider range of possible physical outcomes in terms of regional impacts on the climate. Even if these physical outcomes were known for certain, the economic impacts of climate-related damages would be uncertain and difficult to quantify, not least because the largest damages will occur in the future, and there is considerable disagreement about how future damages should be discounted in terms of present value.

In addition to climate sensitivity and economic impacts, another important source of uncertainty is the cost of technology for reducing emissions. This feeds into policy-making decisions, as well as directly into the costs that companies will face in meeting mitigation requirements. One of the interesting aspects of this source of uncertainty is that both the level of cost uncertainty and the actual cost of the technology are likely to decrease as a result of increased deployment, both of which should act to encourage policy action to promote uptake (Papathanasiou and Anderson, 2000). These various sources of uncertainty are shown schematically in Figure 2 (note: GHG denotes greenhouse gas).

Mitigation policy is affected by uncertainties in climate response, economic impacts and technology costs. Companies are affected by the resulting uncertainty in the mitigation policies, as well as uncertainty in technology cost.

FIGURE 2

Sources of uncertainty

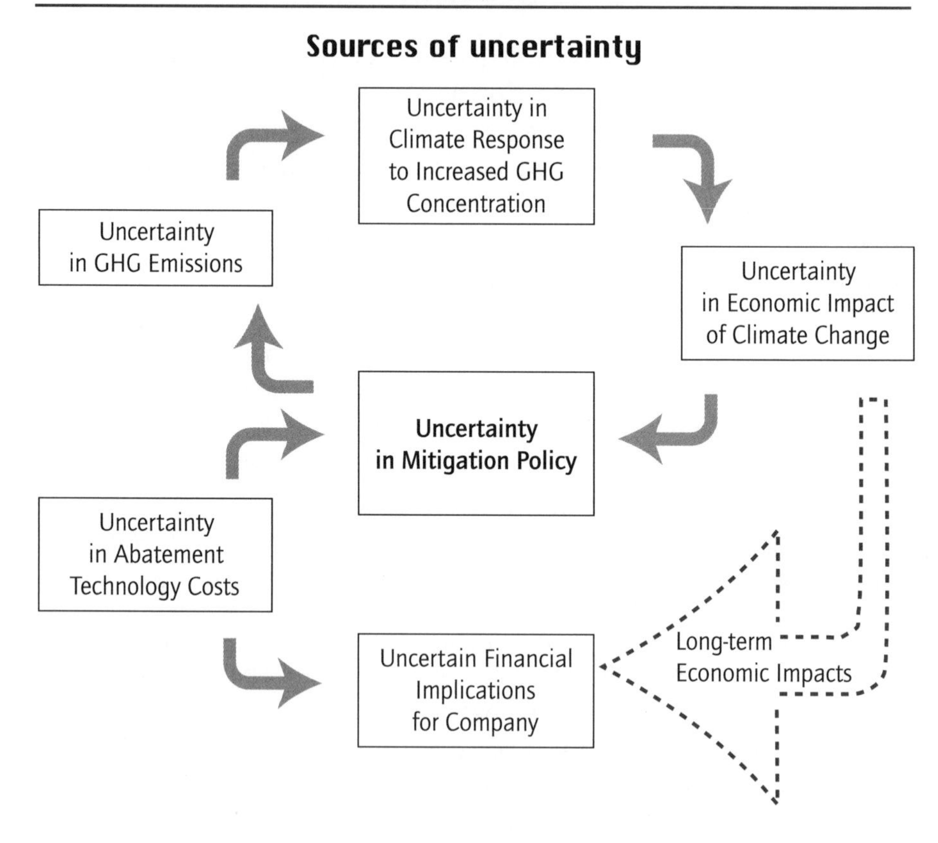

In the long term, there will also be a direct economic impact from physical aspects of climate change.

Various attempts have been made to ascertain the relative contributions of different sources of uncertainty. Nordhaus and Popp (1997) modelled the value of improving knowledge in various aspects of the climate-economic system. Of seven uncertain parameters they modelled, the two most significant uncertainties were the cost of climate change damages and the cost of mitigation, with the climate sensitivity parameter coming third. However, the likelihood of reducing uncertainty associated with climate damages may be limited (Jacoby, 2004). Webster *et al.* (2003) indicate that uncertainties in the emissions profile of greenhouse gases over the coming century are at least as important as uncertainties in the underlying science.

Others argue that there is a value to waiting until some of these inherent uncertainties are resolved before making decisions on restricting emissions of greenhouse gases. This result arises because of the irreversibility of abatement decisions related to long-lived capital stock. This argument says that the expectation of learning more information in the future should lead to higher emissions now in order to avoid a situation where greater abatement actions were taken than proved necessary. A converse argument suggests that the irreversibility of climate damages dominates, and therefore the prospect of learning in the future should lead to lower emissions now (in order to avoid higher-than-expected damages). The ability to learn in the future could indicate either more restrictive or less restrictive policies. Different positions taken in this debate over the role of learning tend to be driven mainly by preconceptions over the cost of reducing emissions: those expecting higher costs tend to favour waiting, those expecting lower costs tend to favour early action (Webster, 2002; Yohe *et al.*, 2004).

Other studies have tried to assess the value of taking early action in the face of these uncertainties by assessing the potential impacts on technology cost, recognising that technology cost may be uncertain, but is in any case expected to reduce as utilisation increases. Papathanasiou and Anderson (2000) argue that technology policies in particular tend to be quite robust in the face of uncertainty due to this positive feedback effect on technology costs. In this sense, kick-starting investment in new technology creates opportunities for "pleasant surprises" that could act as a counterweight to the possible "nasty surprises" coming from climate change impacts. Accounting for emission mitigation efforts also reduces the probability of very high-damage catastrophic events (Frame, 2005).

The existence of low-probability, high-damage outcomes of climate change ("nasty surprises") gives climate damages a skewed probability distribution towards higher damages. Schneider (2000) argues that the significant chance of such nasty surprises should cause a higher level of optimal abatement than if policy was directed simply at the most likely outcome. In other words, policy should take some account of the insurance benefits of reducing the probability of the highest damage outcomes. In principle, this should skew the probability distribution of climate regulation that companies face towards more stringent controls on greenhouse gas emissions.

Such a risk-based approach has been at the heart of the Stern review, which concluded in favour of taking early action to reduce emissions on the basis of a significantly higher cost of damages compared to the cost of abatement (Stern, 2006).

Theoretical approaches to dealing with risk

Businesses routinely deal with many sources of risk and uncertainty in their daily decision-making processes, as illustrated in Table 1. Companies will use their own techniques for assessing these investment risks and deciding how to balance individual project risk against their own strategic concerns. The interaction between company-level decision making and broader financial market treatment of risk is also important.

TABLE 1

Types of risk facing energy sector investments

Category	Types
Economic Risk	Market risk Counterparty credit risk Construction risk Operation risk Macroeconomic risk
Political Risk	Regulatory risk Transfer-of-profit risk Expropriation/nationalisation risk
Legal Risk	Documentation/contract risk Jurisdictional risk

Source: Adapted from *World Energy Investment Outlook* (IEA, 2003).

This section provides a brief overview of the theory of how risk can be incorporated into investment appraisal. This sets the scene for introducing the analytical framework used in this study described in Chapters 2 and 3. A discussion of the different ways in which companies manage risk in the real world is deferred to Chapter 4.

Incorporating risk into NPV calculations

Project appraisal is concerned with the assessment of the value of investing capital today in return for an income stream in the future. There are several different ways of representing the financial prospects for a proposed project, including simple paybacks, internal rates of return and return on capital employed; each of which has its own merits and drawbacks, and each of which is used in different ways by companies in decision making. It is generally agreed, however, that the most rigorous way to carry out project appraisal is to calculate the *present value* of the future income stream, and subtract the capital expenditure required to give the *net present value* (NPV). If the NPV is positive, the project should be profitable. If a choice is to be made between two projects, the one with the higher NPV should be preferred.

Present value takes account of the time value of money. It is calculated by discounting future earnings by some amount that reflects the investor's preference for holding money today rather then earning money in the future. Ignoring uncertainty for the moment and assuming that money for the project can be borrowed at a risk-free rate *(r)*, a single up-front capital investment cost *(I)* and annual cash inflows (c_t) in period *(t)*, the NPV is calculated as:

$$NPV = \sum_{t=1}^{T}\left[\frac{c_t}{(1+r)^t} - \frac{I}{(1+r)^t}\right]$$

As long as the cash flow elements (c_t) were certain, the positive NPV criterion for investment would be the correct one.

In the real world, many different elements of the project finances will be uncertain. For example, the investment costs may be different from expected due to technical risks (the risk of cost overruns when the project is being built may be particularly important in the early adoption of new technologies). All power generation projects will have uncertainty relating to the cash inflows because of changes in either revenues or operating costs. These reflect underlying uncertainties in primary variables such as the costs of primary fuels and price and demand fluctuations in the electricity markets.

How can these risks be incorporated into the NPV calculation? In principle, each uncertain element of the cash inflow *(c_t)* should be replaced with a *certainty-equivalent* amount *($\hat{c}_t$)*. The value of certainty equivalent *($\hat{c}_t$)* in a specified period *(t)*, in this case a given year, is chosen such that it has the same *present value (PV)* as the uncertain cash inflow when they are discounted at the appropriate (risk-adjusted) rate. Hence, incorporating risks in the NPV is calculated as:

$$PV = \frac{\hat{c}_t}{(1+r)^t} = \frac{E(c_t)}{(1+k)^t}$$

Where E*(c_t)* is the expected (mean) value of the uncertain cash inflow, and k is the opportunity cost of capital (or the risk-adjusted discount rate) for projects of that class of risk. The appropriate value for k will be discussed further below. Then the NPV under uncertainty can be written as:

$$NPV = \sum_{t=1}^{T} \left[\frac{\hat{c}_t}{(1+r)^t} - \frac{I}{(1+r)^t} \right]$$

This certainty-equivalent approach disaggregates the effects of the time value of money under certainty from the effects of risk. Equivalently, one may define a risk premium as the expected value of the cash flow in a given year minus the certainty equivalent cash inflow. This risk premium should reflect the overall market risk premium for that class of project. In general, the cash inflows in each period may be subject to a different level of risk, requiring a different risk premium to be used for each period. This would be the case if project uncertainty were resolved in a "lumpy" manner rather than being gradually resolved in a smooth way over time. This is important in the context of policy uncertainty, which is likely to be resolved through discrete information events such as the announcement of a change in policy.

It may also be important for different elements of the cash flow to be discounted using different risk premiums (for example, gas prices may be deemed more risky than coal prices). However, it is difficult to determine the correct adjustments that should be made to the different elements of the cash flow in each period, and the risk premium for a "new" risk such as climate change policy uncertainty may be difficult to determine.

In practice, a simplifying assumption is often made in order to allow both the time value of money and the project risks to be represented by a single risk-adjusted discount rate for the project as a whole; whereby the calculations appear as:

$$NPV = \sum_{t=1}^{T} \left[\frac{E(c_t)}{(1+k)^t} - \frac{I}{(1+k)^t} \right]$$

This simplified treatment of risk in the NPV calculation is widespread amongst firms and in macroeconomic energy models in practice. For projects where the risks are well known and resolved smoothly over time, it is appropriate. Projects in the same line of business as the firm and having the same risk characteristics do not affect the firm's total riskiness, so that k should be the same as the company's average cost of capital. For projects in a different risk class from the company as a whole, a different discount rate should be used, one that corresponds to the opportunity cost of capital for other projects in that risk class (*e.g.* those being less risky attracting a lower discount rate, those being more risky attracting a higher discount rate).

The assumption of a single constant discount rate for the whole project implicitly assumes that the risk per period of the project is constant. This in turn depends on an assumption that uncertainty is resolved in a smooth manner. This assumption may break down if the uncertainty is likely to be more "lumpy", with large amounts of information being revealed at certain points in the project. If the company has flexibility to respond to uncertain events by adapting their behaviour (*e.g.* by changing their level of investment in subsequent phases of the project), this can also affect exposure to risk and invalidate the assumption of a fixed discount rate.

Real options approaches provide a way to overcome these difficulties by treating uncertain events and flexible management responses explicitly. Before describing the real options framework used in this study however, it is important to understand how risk is handled at the broader level of the economy as a whole in order to place these arguments in a wider context.

CAPM and market price of risk

The risk-adjusted NPV calculation requires the use of a discount rate that suitably reflects the risk of the project being considered. How can the correct risk adjustment be determined? An intuitive answer is that the project risk should be determined as a function of the variance (spread) of the cash flows. For example, a project facing greater uncertainty would be expected to have a wider spread of possible outcomes than one with less uncertainty, and so should be discounted more strongly. However, this does not take into account the role that wider financial markets play in hedging project-specific risks.

Strictly speaking, the discount rate should equal the opportunity cost of capital for the project, which is how much the market would require in terms of return in order to take on a project with that class of risk. So how does the market evaluate risk for the project?

It is a fundamental principle of equity markets that investors can reduce their risk while expecting to earn the same return by spreading their investment over a portfolio of different stocks. The reason for this is that returns from different stocks will vary over time, but do not always move in the same direction as they fluctuate – positive moves in returns from some stocks will offset negative moves in other stocks making the average fluctuation of the portfolio returns lower than the fluctuations in the individual stocks. The extent to which the fluctuations cancel each other out depends on the correlation between the returns of the stocks in the portfolio.

So from the point of view of an average shareholder who can choose to invest in any of the stocks in the market, the relevant factor when considering the riskiness of a new project is not simply the total variance in returns on the project, but the extent to which that project contributes to the riskiness of the whole portfolio. The overall level of risk that the project presents to a fully diversified shareholder is therefore proportional not only to the variance in returns of the project itself, but also to the correlation between the project's returns and the returns of the market as a whole.

This principle is captured in the *capital asset pricing model* (CAPM). This expresses the expected rate of return for a project in terms of an overall market premium, which reflects the risk of the equity market in general, and a project-specific beta factor (β) that incorporates the variance of the project cash flow and the correlation between this variance and the overall market. The CAPM can be expressed as:

$$k_j = r + \beta \times [m - r]$$

Where k is the expected (required) market return for the project, *r* is the risk-free rate (*e.g.* as represented by gilt bonds), and *m* is the expected return from the whole equity market. The factor *[m - r]* represents the overall market risk premium for holding equities rather than gilt bonds. The *beta* factor *(β)* is usually expressed in terms of the *covariance* between the projects return and the market return. This means that it is not just the variance of the project's returns that matters – the *covariance* measures both the variability of the projects return and the correlation of this variability with market fluctuations. This means that project risk may be considered to fall into one of two categories:

- *Systematic* risks (otherwise known as market or non-diversifiable risks) arise from variability in return that is driven by economy-wide factors which affect the market as a whole, and cannot therefore be diversified away.
- *Specific* risks (or diversifiable risks) are unique to individual companies and relate to variability driven by factors that are not related to general economic factors, such that they can be diversified away by holding an appropriately diverse portfolio.

Hence, according to the CAPM model, only the systematic risks should feature in the valuation of risk when determining the appropriate risk-adjusted discount rate to use in the project appraisal. The CAPM model provides a market equilibrium view of how risks should be treated by financial markets, and therefore how these should be incorporated into a company's treatment of risk in their project evaluations. In principle, project *beta* factors should be calculated based on a probability distribution of returns and a view of the correlation of these distributions with market-wide variability.

In practice, there are limits to the absolute levels of risk (either systematic or specific) that companies can take before their credit risk is affected, which would drive up their cost of debt. Companies may therefore be more inclined to diversify their assets than the pure CAPM model would suggest. It may also be difficult to establish a suitable beta factor, especially if the risk changes in different periods of the project (as would be the case, for example, if uncertainty is resolved in discrete events rather than continuously at a constant rate throughout the project). Typically, a company may have a feel for its overall beta factor value (which may be similar for the sector as a whole), but would not usually try to calculate a *beta* factor for each new project undertaken. These practical considerations are discussed further in Chapter 4.

Again, real options approaches provide a useful way as described in the next section to quantify climate change policy uncertainty. It should be noted that the analysis provided by real options is an extension of the standard NPV analysis. The project appraisals we look at using real options therefore still sit within an overall market equilibrium context (*i.e.* the use of real options does not alter the appropriateness of the CAPM as a model of the broader financial markets' response to project risks).

Rationale for the use of real options

Real options mean different things to different people. For some, real options have connotations associated with the collapse of Enron[3], which has led some companies to move away from their formal use. Nevertheless, the methodology itself remains a robust way of incorporating risk into project appraisal, and in the context of this study, provides a powerful tool for policy analysis. The modelling technique used to evaluate option values is known as stochastic optimisation. In this book, we use the terms real options and stochastic optimisation interchangeably.

Real options approaches explicitly incorporate individual elements of risk into the cash- flow calculation, taking into account management's flexibility to adjust their behaviour, as the uncertainties get resolved. This ability to analyse explicitly

3. Enron Corporation was an American energy company based in Houston, Texas. Before its bankruptcy in late 2001, Enron employed around 21 000 people (McLean & Elkind, 2003) and was one of the world's leading electricity, natural gas, pulp and paper and communications companies, with claimed revenues of USD111 billion in 2000. *Fortune* named Enron «America's Most Innovative Company» for six consecutive years. Enron achieved infamy at the end of 2001 when it was revealed that its reported financial condition was sustained mostly by institutionalised, systematic and creatively planned accounting fraud. Enron has since become a popular symbol of wilful corporate fraud and corruption. The lawsuit against Enron's directors, following the scandal, was notable in that the directors settled the suit by paying very significant amounts of money personally. In addition, the scandal caused the dissolution of the Arthur Andersen accounting firm, which had effects on the wider business world, as described in more detail below. Enron still exists as a shell corporation (without assets). It emerged from bankruptcy in November of 2004 after one of the biggest and most complex bankruptcy cases in U.S. history. On September 7, 2006, Enron sold Prisma Energy International Incorporated, its last remaining business, to Ashmore Energy International Ltd. According to the final restructuring plan submitted to bankruptcy court, Enron will be dissolved at the conclusion of the restructuring process.

the effect of a particular source of uncertainty on an investment decision is precisely the reason why a real options approach has been taken in this study. The purpose of using these techniques is not to claim that they are superior to other forms of risk management (although such claims can be made - see Trigiorgis, 1996). Real options approaches provide a powerful tool for giving insights into the questions that motivated this study, namely *"does climate change policy uncertainty pose a significant risk to power sector investments, and if so, how could policy design be improved to reduce these risks"?*

Real options theory (Dixit, 1994; Trigeorgis, 1996) borrows from the theory of pricing financial options. Financial options give the holder the right, but not the obligation to buy a stock at some specified future maturity date at some specified price (the "strike price"). Because of the right without obligation to exercise, options have a non-symmetric risk profile: if the price of the stock at maturity turns out to be lower than the strike price, then the holder will not exercise, the option is worth zero, and no loss is incurred. If the price of the stock at maturity turns out to be higher than the strike price, then the holder will exercise the option, buying at the lower strike price with the ability to make an immediate profit. As soon as it is exercised, the option is used up, and has no further value. In order to find a counter party willing to sell such an asymmetrical option, an initial payment must be made for the option. The value of a financial option depends, amongst other things, on the volatility of the underlying stock price. If there is no volatility (*i.e.* full certainty), then options are worthless, since there is no value to hedging downside risk. Conversely, option values increase under conditions of greater volatility.

Real options theory draws an analogy between financial options and investments in physical assets. Consider a company facing a decision to invest in a new plant, and assume it has the freedom to choose whether or not to proceed with the investment at any time over a specified period, for example, five years. If the investment is made, it is more-or-less irreversible, since the plant cannot be resold without losing considerable value. If during those five years conditions justify the investment, then the investment will be made, and a positive return may be made (analogous to the returns gained by exercising the financial option). If, on the contrary, conditions are not favourable, the investment will not be made and no loss will be incurred (analogous to the financial option not being exercised). The "option" to invest therefore has the same characteristics as a financial option - once exercised it is binding and irreversible, and it has an asymmetric pay-off, which gives it an inherent value. Therefore, the "option" to invest in a new plant has a value of its own, over and above the financial pay-off of the project itself. The expected revenues from a proposed new project will therefore need to cover not only the capital costs of the project, but also this additional option value. This increase in required project returns effectively acts like a risk premium associated with the uncertainty in project revenues.

The reason why a risk premium needs to be incorporated into the investment decision is clear when one considers that the act of investment moves the company from a position of flexibility to a position where having made the investment it is locked in to a certain irreversible course of action. The company will need to be recompensed for giving up this flexibility, which is the source for the "option value" effect. In the absence of uncertainty, the option value reduces to zero, since there is no value to maintaining flexibility when the future is known for sure. In this case, the pay-off from the project only has to cover the initial capital cost, replicating the standard positive NPV criterion.

Real options can be used in a formal way to value investment opportunities available to companies as long as the risk profile for the project is well understood and is matched by some tradable security on financial markets. However, in this analysis, real options theory is used in a less formal way; we use a subjective Bayesian probability distribution for future changes in climate policy, which cannot really be tested against a traded security. Our approach is therefore similar to simulation models and decision-tree analyses that make explicit assumptions about the likelihood of occurrence of different future events, and values of key variables such as carbon and fuel prices. Such approaches have become quite widespread in the analysis of market, technical and regulatory risks (see for example, EPRI, 1999; Epaulard, 2000; Ishii, 2004; Kiriyama, 2004; Rothwell, 2006). The real options framework used in the study is described in more detail in the next chapter.

APPROACH TO QUANTIFYING UNCERTAINTY

As described in Chapter 1, real options theory is an extension of standard project appraisal methods, adding the ability to explicitly model the effect of individual sources of uncertainty, and accounting for the flexibility that managers often have over the timing of their investment when faced with uncertain future cash flows.

A way to understand the approach is to realise that investors, faced with a risky irreversible decision, will value the opportunity to gain additional information about likely future conditions affecting the project, thereby reducing uncertainty. This could mean investing in additional research for example, or more relevantly for our work here, delaying investment until the uncertainty has been partially resolved.

When cash flows of a project are uncertain, the value of waiting for additional information depends on the extent to which the uncertainty affects the cash flow, how far in the future the uncertain event is, and the likely quality of the information that will be gained by waiting (*i.e.* the extent to which the uncertainty will be resolved).

A project may still go ahead without delay if the project value under current conditions (the opportunity cost of waiting) is sufficiently high that it exceeds this value of waiting. This is described in Figure 3.

Uncertainty can be conceptualised as an anticipated price shock or an information event (*e.g.* introduction of a major new climate change policy) at some time (T_p) in the future. This could affect a project's cash flow either adversely or favourably. In Figure 3 Case A, the company facing this uncertain cash flow has to choose whether or not to invest in the project - it does not have the option to wait. The expected "best guess" (central orange line) is that the project will continue to be profitable, so that the project satisfies the normal investment rule (*i.e.* the gross margin is greater than capital cost) justifying immediate investment.

In Figure 3 Case B, the company has the opportunity to wait the expected time (T_p) before making the investment. This allows it to avoid the potential loss that might occur if conditions turn out worse than expected (shown as a red dashed area). Waiting could lead to a greater return on investment – the new expected gross margin from the project would be higher than the original expected gross margin without the option of waiting – but revenues from the project would only accrue after the expected time (T_p) if the project does go ahead. It would be rational to invest prior to the expected time (T_p) only if this value of waiting is overcome by the opportunity cost of waiting (*i.e.* the income forgone due to delaying the investment). In order to trigger immediate investment, the expected gross margin of the project would need exceed some threshold level that makes the opportunity cost of waiting greater than the value of waiting. This threshold depends on the length of time before the information event or policy change affects change (T_p), the size of the anticipated price shock and the discount rate.

Understanding the possible value of waiting to invest leads to a revision of the standard investment rule. The standard discounted cash flow (DCF) analysis would indicate that investment should proceed if the revenues exceed the costs (when these costs and revenues are discounted back to current values). Incorporating flexibility on the timing of investment would give the following revised rule:

Traditional investment rule

Invest now if the discounted revenues exceed the discounted costs.

Revised investment rule

Invest now only if the discounted revenues exceed the discounted costs by a margin sufficient to overcome the value of waiting.

The results we present later in the report are presented in terms of the additional investment threshold required to overcome this value of waiting in order to justify immediate investment in the project. This approach allows us to look explicitly at the effect of different sources of uncertainty.

Such approaches have been used before to model the effects of regulatory uncertainty on the power sector, and have been useful in helping to explain deviations from investment behaviour that might have been expected based on a simple NPV model (Ishii, 2004; EPRI, 1999).

FIGURE 3

A conceptual framework for understanding the value of waiting

Case A : "Now or never" investment option at t = 0

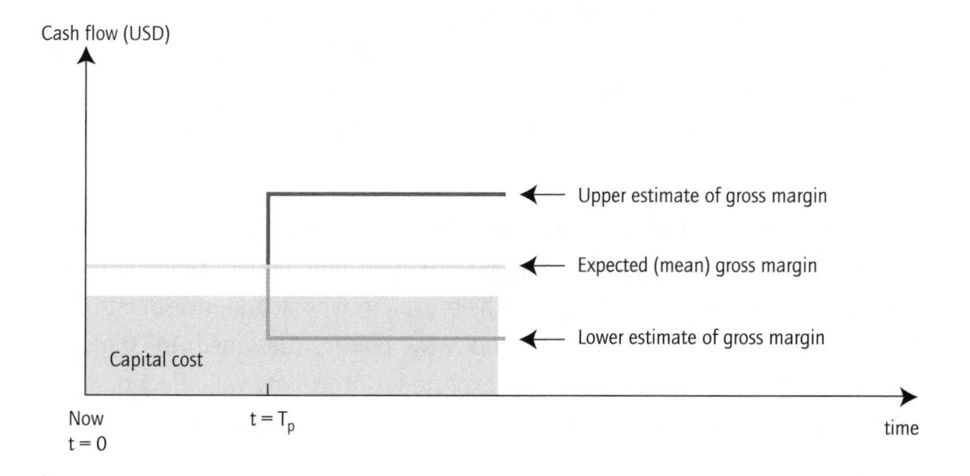

Case B : Company has the option to wait until after $t = T_p$, the expected time of some policy change that affects the investment

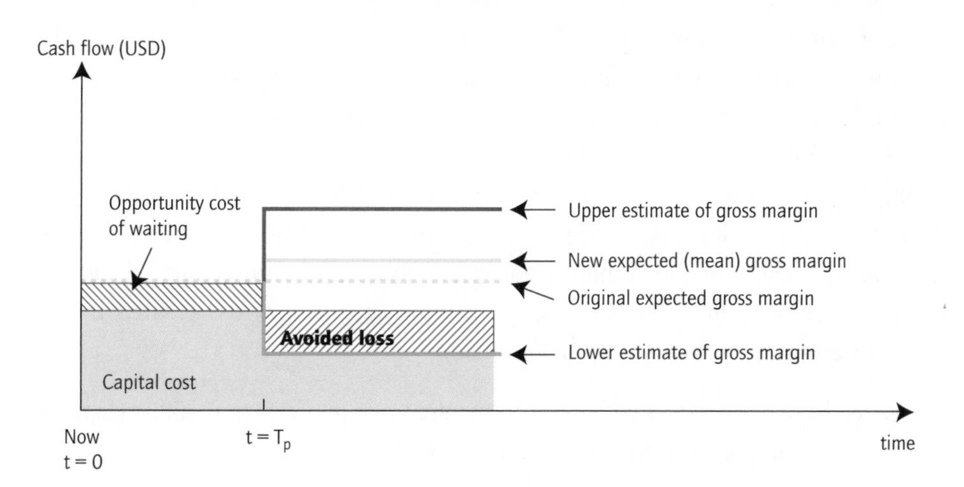

Relevance to policy makers

When considering a proposed new policy, climate change and energy policy makers need to make some estimate of the level and type of investment that will be made by companies under different economic conditions, and how this investment behaviour might be changed by the new policy. The analyses used to inform policy decisions usually take account of many different economic factors (including prices), but usually do not take explicit account of risk. The consideration of risk can significantly alter investment decisions. This means that companies may not respond to government policies in the expected manner.

A stochastic optimisation approach may indicate that the price of carbon required to trigger investment in low-emitting technologies may be substantially higher than previously expected. It may also help explain why actual investment rates tend to differ from expectations and why policies designed to stimulate investment may be less effective than expected. Uncertainty will affect different technologies to different extents, and so the presence of uncertainty may lead to unexpected trends in technology uptake.

Carbon price is used in this report as a shorthand way of representing the effective financial value that a company will place on avoiding emitting greenhouse gases. The company might experience this price directly if it is in an emissions trading scheme where the price is derived through market trading activities, or if there is a carbon tax where the price of carbon is explicitly stated. Alternatively, the company might experience the carbon price indirectly as the internal cost of meeting some regulatory requirement or investment incentive (*e.g.* emissions standards, technology quotas and/or grants). Either way, the price perceived by the company is created by policy – it is assumed that there is no inherent value to the company in avoiding greenhouse gas emissions except insofar as it helps meet regulatory requirements. Since the carbon price is essentially created by government policy, the uncertainty surrounding this price is also created by government policy – or more specifically, it is created by the uncertainty surrounding government policy.

We might re-phrase the real options investment rule in a way that makes clear the role of and importance to climate policy makers:

Revised investment rule (from the perspective of policy makers)

Investment will only proceed as expected if the required investment conditions can be guaranteed to remain in place forever. Any uncertainty over future conditions that affect the project cash flow will cause investment behaviour to deviate from these expectations. If the uncertainties are large, these deviations may also be large.

As well as providing insight into what these deviations from expectations might be, this approach provides a useful tool for analysing the likely effects of different policy designs. Another advantage of casting the issue of uncertainty in terms of options value is that it takes the discussion beyond the normal policy debate about investment "barriers". Barriers are generally hard to quantify, whereas options effects can be more readily quantified. The distinction is also important from a policy point of view. Whereas the presence of barriers indicates some kind of market failure, in this formulation behaviour is seen as rational, so a different policy response may be warranted. Implications of the analysis for policy makers are discussed further in Chapter 5.

How do we represent uncertainty?

The analysis looks at two important sources of uncertainty–climate policy uncertainty and fuel price uncertainty–to provide some comparison and context. Clearly, there will be many other uncertainties and risks for companies facing new investments that are not explicitly modelled here. Some qualitative discussion of these other risks is given in Chapter 4.

In this study, all uncertainty relating to climate change policy is expressed through the carbon price. For carbon prices, we can distinguish between three different types of variation:

- Short-term (less than one year) volatility. This occurs where prices fluctuate quite rapidly according to conditions in the market. We make the simplifying assumption that this type of rapid price variation does not significantly alter the investment decision, since medium to long-term expectations of prices drive investment. For most of the results presented in this report, we ignore the effects of short-term volatility, although we do revisit its potential to increase project value in the face of operational flexibility.

- Longer-term (greater than one year) price uncertainty. This type of variation is used to represent underlying uncertainties (*e.g.* those relating to weather, technology costs and other market conditions). We model price variations with a random walk (geometric Brownian motion) price process with annual standard deviation of ± 7.75%. This means that expectations of future prices may drift up or down by this amount each year. The total range (standard deviation) of prices after 15 years of random walk would be ± 30%. This is chosen to match the range of gas price uncertainty used in the model.
- Climate policy uncertainty. In addition to the underlying uncertainty described above, we model policy uncertainty as a discreet jump in price at some known time in the future. This represents an "information event" or price shock resulting from a policy announcement such as might arise at the beginning of a new allocation period in an emissions trading scheme, or the announcement of a new tax rate, technology standard or other regulatory approach. We model this with a jump in carbon prices in the range ± 100% with a flat probability distribution within this range. The year in which the jump occurs can be varied – we compare results where the jump occurs after 5 years versus a case where the jump occurs after 10 years.

The second and third types of uncertainty are illustrated in Figure 4. The expected (mean) value for CO_2 price is allowed to vary depending on the technology being considered, and is therefore not shown in the figure. The value is chosen to make the technology being considered financially viable.

The IEA's model also includes fuel and electricity price uncertainty. Following the approach described by Pindyck (1999), fluctuating fuel prices are treated as a combination of volatility (modelled as a short-run mean reversion process), with a mean which itself is uncertain and can drift according to a random walk process. In most of our results, we ignore the short-term volatility, and simply model fuel price primarily through a long-run random walk process similar to that described in the second bullet point above (*i.e.* a geometric Brownian motion price process). Gas and oil prices are modelled with an annual standard deviation of ± 7.75% and coal with a ± 1.8% standard deviation. This gives a standard deviation from the expected mean after 15 years of ± 30% for gas and oil

FIGURE 4

Presentation of climate policy uncertainty

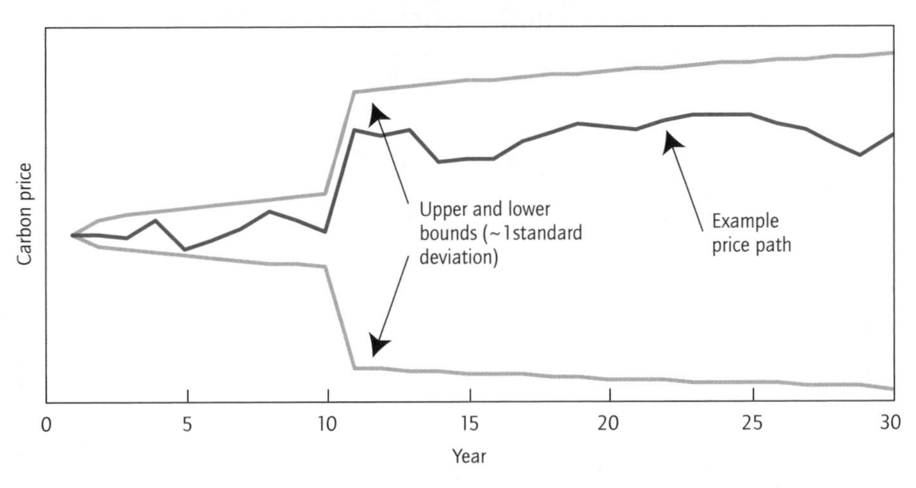

Climate policy uncertainty is represented through price uncertainty, as a possible jump in price (either up or down) at some known time in the future. Underlying uncertainty in price is treated as an additional random walk process, which would give a range of ±30% over 15 years in the absence of any price shock.

and ± 7% for coal, approximately in line with the International Energy Agency's high and low price scenarios (IEA, 2004a). The upper and lower expected price bounds (1.0 standard deviation) as a function of time and an example random walk price path is shown in Figure 5. The expected (mean) price level for gas is allowed to alter slightly between runs in order to help balance the expected cash flows for the various technologies being considered, but is generally in the range of USD 5.2-5.6/GJ (USD 0.55 to 0.59 per therm) although this is not shown in the figure. The expected (mean) price for coal is USD 1.9/GJ.

As a sensitivity case in some of our results in Chapter 3, we introduce short-term volatility in addition to this random walk, but in general, we assume that only the longer-term uncertainty in fuel price will affect investment decisions. We do not model price shocks for fuel, although in principle these could be incorporated into the model.

The expected (mean) fuel prices are assumed to be flat rather than growing over time. This is because we want to model the effect of uncertainty on the timing of investment, and it is easier to separate this effect out if the underlying economic case does not change over time. If expected fuel prices are allowed to vary over

FIGURE 5

Modelling uncertainty in fuel prices as a random walk process

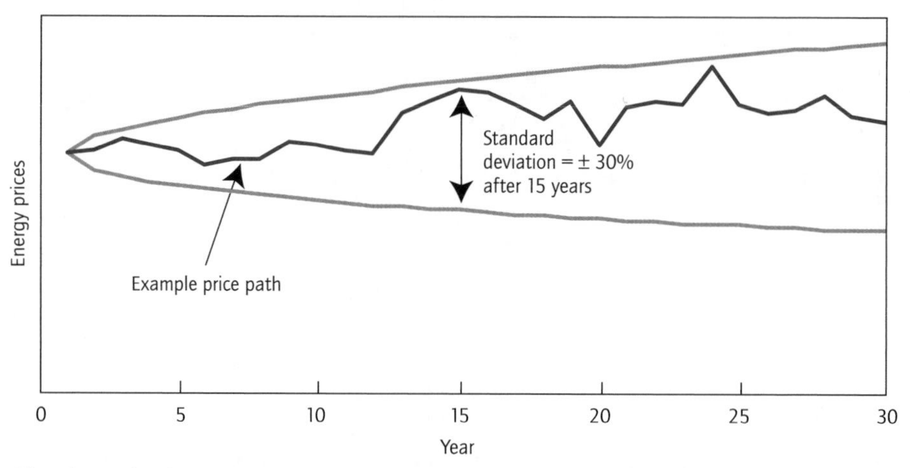

Oil and gas prices have a standard deviation from the expected mean price of ±30% after 15 years, and ±7% for coal.

time, this may lead to a value of waiting, which arises from changes in the economic case for the project (*e.g.* a fuel switch option may be more attractive under expected future prices than under current prices). This effect would need to be extracted from the overall value of waiting derived from the stochastic optimisation model in order to derive the risk premium associated with price uncertainty. Since the results presented in this book are essentially for illustrative purposes, we have simplified the analysis by assuming that the economic case does not change over time. However, the approach could be readily adapted to take account of more realistic fuel price scenarios.

It is assumed that CO_2 and gas price fluctuations are at least partly correlated - different sensitivity cases are presented with different values for the correlation factor. Electricity prices are derived from the short-run marginal cost of either gas or coal generation (including fuel costs, variable operating and maintenance costs and CO_2 costs), with an additional spark spread value, which is also assumed to be stochastic so as to approximately replicate the ±30% standard deviation when gas is on the margin. Since electricity prices also incorporate CO_2 prices, they will be subject to a jump in price similar in nature to that shown in Figure 4.

How does the model work?[4]

The key drivers of investment in the power sector will be expectations of future prices of electricity, fuel, technical and operating costs, as well as environmental and other legislative requirements (including CO_2 emission costs). These elements are incorporated into a cash flow model for the technology being considered. The cash flow model gives a figure for the expected present value of the future cash flow, which could be compared with the capital cost of the project. In simple terms (not accounting for project risk), the project would be considered financially viable if the net present value (NPV) of the project is positive.[5]

The model then incorporates randomised ("stochastic") price paths similar to those shown in Figure 4 and Figure 5 for the main elements of the cash flow such as electricity, gas, coal and CO_2 prices. The cash flow model extends for the technical duration of the plant (*e.g.* assumed to be 25 years for a gas plant, 40 years for a coal plant). Certain technical assumptions are made in the model, for example, generation efficiency, emission factors and operating load, which determine how the different input prices are translated into revenues and costs in the cash flow. These technical assumptions are shown in Appendix 1. The price paths therefore represent an "input" to the model. The model then calculates the total present value of lifetime cash flows. This measures the total revenues minus the total operating costs adjusted for tax and build time, which is then discounted to give the total present value over the expected lifetime of the plant. A schematic for the cash flow is given in Figure 6. For simplicity, this final output from the cash flow is referred to from here on as the gross margin.

The model uses a Monte Carlo method to build up a probability distribution for the gross margin by running the cash flow many times, each time using a different price path for the uncertain variables. Over several thousand runs, a more-or-less complete picture develops of the possible different combinations of input prices for each year of the lifetime of the plant within the bounds illustrated in Figure 4 and Figure 5.

The option to invest is incorporated by allowing the model to make an irreversible "switch" between different cash flow streams in any given year. In order to make

4. A detailed technical description of the model is available in an IEA working paper (Yang and Blyth, 2007).

5. NPV = Present value of future revenues less present value of future fuel costs, less present value of future operating costs, less present value of future environmental costs, and less capital costs.

FIGURE 6

Structure of cash flow model

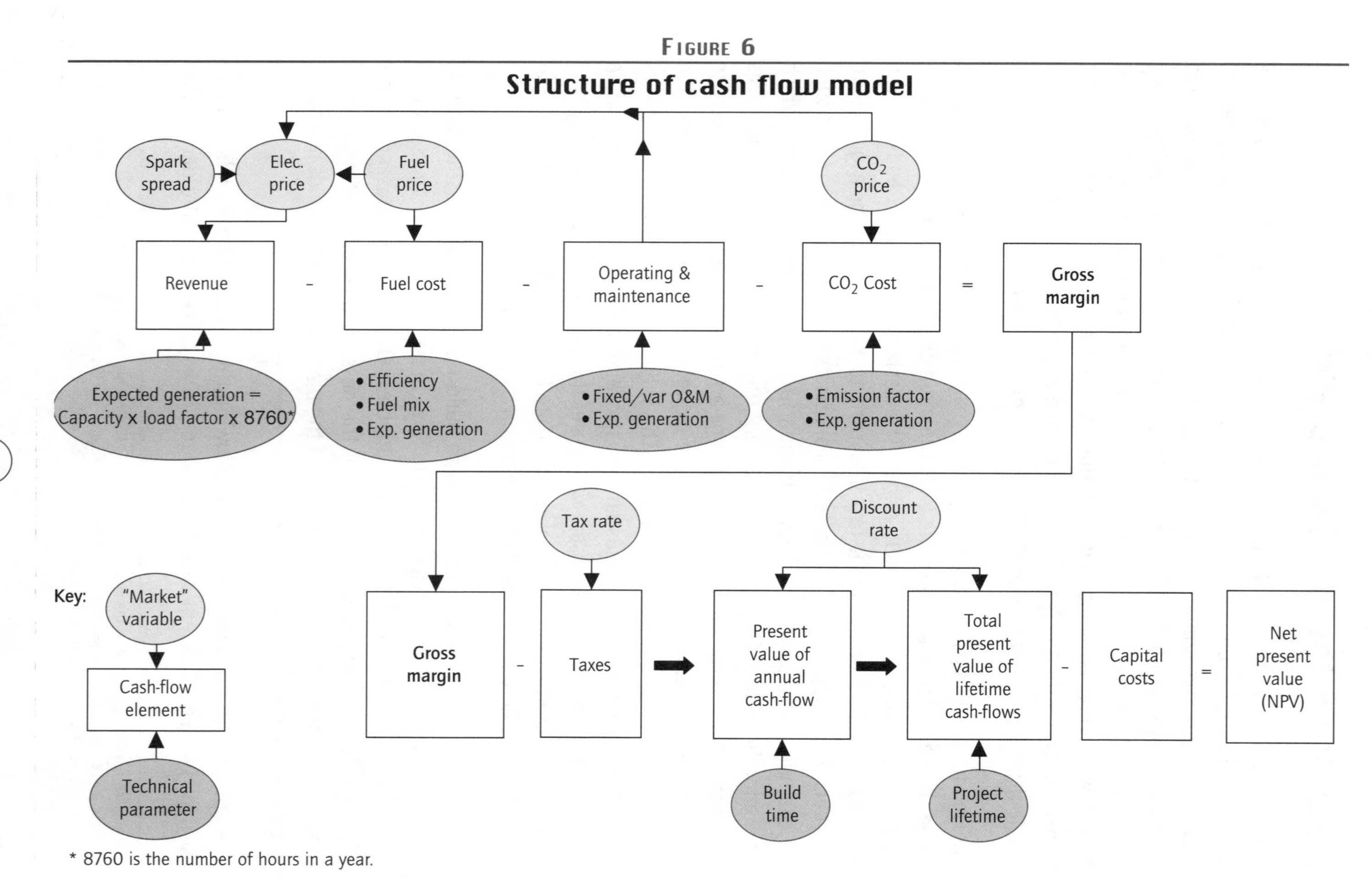

* 8760 is the number of hours in a year.

the switch, the capital cost of the investment has to be paid. Therefore, each year the model can choose between the existing cash flow (effectively a "do nothing" option), and another cash flow which represents the expected gross margin of the technology being considered (an "invest" option). This choice to switch is made on the basis of whether the expected pay-off from making the investment (taking account of capital costs) is better than the expected pay-off from sticking with the existing cash flow. The difference between this calculation and a normal comparison of DCFs of an investment opportunity is that the valuation of the "do nothing" option includes the possibility that the investment will be made later. The model is able to optimise the timing of investment by building up an investment rule that takes account of expected future events (*e.g.* price shocks or policy interventions), and comparing current conditions against this investment rule.

The investment rule is established by running the simulation of prices "backwards"; that is starting from the end-date and working back towards year one, and in each year comparing the expected value of the "do nothing" and "invest" options, where the "do nothing" option takes account of optimised values in future years that have been calculated in previous steps. This is a dynamic programming approach as described in Dixit and Pindyck (1994). The Monte Carlo and stochastic optimisation elements of the model are based on semi-commercial software used by the Electric Power Research Institute of the USA (EPRI), with the cash-flow calculations being performed in Microsoft Excel.

The cash flow model used contains assumptions about technology (*e.g.* generation efficiency, capital costs, operating costs, fuel type, emission factors and load factors); financial assumptions (*e.g.* discount rate and tax rates); and commodity prices (electricity, gas, coal, oil and CO_2). The cash flow model has to make an assumption about the linkages between these different commodity prices. In general, all prices in a real options model tend to be exogenous (*i.e.* externally specified). However, the model does allow a functional relationship between the prices, since price correlations between electricity, gas and carbon are important.

Interpretation of options value

In contrast to some other real options studies, we do not use a typical scenario approach to look at the cost-effectiveness of different investments. Instead, in this study, we frame the problem slightly differently. The main topic of this study is to understand the difference that uncertainty can make to an investment decision – in particular, the extent to which it raises the investment threshold.

It is useful to introduce another simple graphical representation of how options values increase the investment threshold. Suppose a company is considering investing in a project for which the gross margin *(M)* is uncertain and could take a range of values with an expected (mean) gross margin, (represented as *E(M)*). What is the value (represented as *F(M)* in the calculation) of this option to invest? Figure 7 shows how the value of an investment option (noted as *F(V)* in the calculation) increases in relation to the expected value of the revenue (written as *E(V)* in the calculation) from a project with capital costs *(I)*. The straight line represents the standard investment rule under certainty: if *E(M)* is greater than I the project is worth *E(M)-I*, but if the *E(M)* is less than *I* then the option to invest is worthless.

The situation under uncertainty is represented by the curved line. At low values of the expected gross margin, the option value *F(M)* is positive even when *E(M)* is lower than the initial capital outlay *I*. This is because under uncertainty, circumstances could change favourably, making the project viable in the future. In these circumstances, the company will want to retain the option to invest, even if it means paying to do so (*e.g.* expenditure might be required for licensing, rent, research and development). As well, when the expected gross margin is at or slightly above the initial capital outlay I, the value of the investment option is greater than the net present value of the project *E(M)-I* as indicated by the fact that the curved line lies above the straight line. This means that the company would rationally hold on to the investment option rather than make the investment itself. Not until the expected gross margin *E(M)* exceeds a further threshold *T* does the net present value of the project equal the value of the investment option *F(M)*, thereby justifying investment. At this threshold value, the curved line meets the straight line, representing the new optimal investment point. Above this point, the value of the project reverts to the normal NPV rule such that *F(M)=E(M)-I*.

The model used in this study provides a figure for this investment threshold *T*, which is the quantity that we represent in our results. The size of the investment threshold depends on the nature of the uncertainty, the major elements of which are captured in the conceptual framework described in Figure 3.

FIGURE 7

The value of the option to invest under uncertainty may be worth more than the NPV of the project

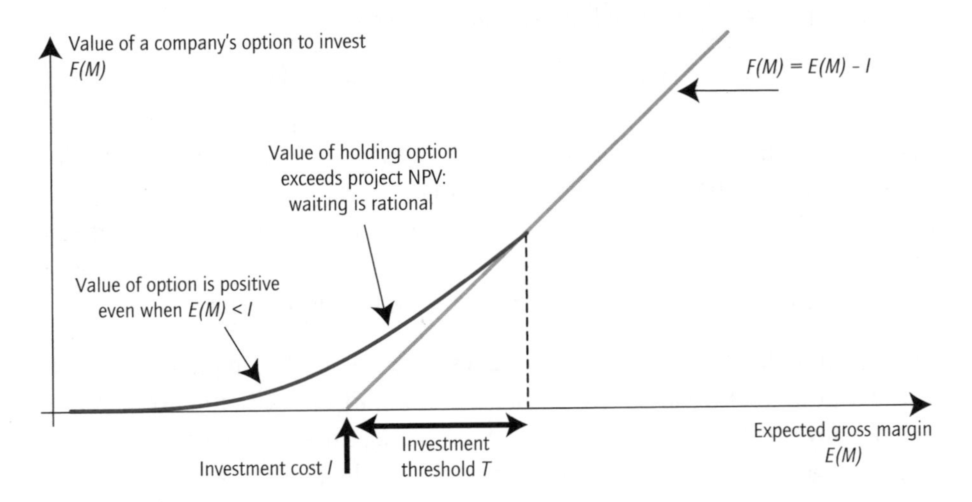

Because we are carrying out a generic analysis aiming to be as relevant as possible to all IEA member countries, we are primarily interested in what the level of this investment threshold is for a given level of uncertainty, not whether the threshold is exceeded under any specific price scenario. For this reason, we have kept the price assumptions used in the model very simple in order to facilitate understanding of what affects the investment threshold. Throughout the study, we try to distinguish between factors that affect the profitability of a project (as calculated using a normal discounted cash flow) and factors that affect the additional investment threshold *T* caused by uncertainty.

In order to calculate the size of the investment threshold using our model, we need to set up our project cash flow so that the expected gross margin is within a reasonable range of this investment threshold. This means making price assumptions in the model that may not match current prices. For example, a significantly higher carbon price is required than today's prices to make carbon capture and storage cost-effective at current technology costs. In our model, we simply raise carbon prices to the required level. This is not really a scenario. We do not claim that these prices are necessarily realistic, just that they are illustrative of what the prices would need to be in order for the project to be considered viable given the assumptions we have made about capital and operating costs and other parameters in the model. The problem is set up to say that, for example, if you want people to consider CCS technology, this is what carbon prices would have to be, and if you take into account carbon price uncertainty, then the required prices would be even higher.

We express the threshold T in the same units as capital cost I so that a comparison can be made in percentage terms. For example, if the results of the model show that the threshold level is 50%, this means that the gross margin for the project would have to exceed capital costs by 50% (*i.e.* so that the total gross margin is 150% of capital cost). As will be seen, this additional threshold can become quite large – in some situations, the gross margin would have to be double the capital costs in order to trigger immediate investment.

It is worth noting that the investment thresholds we are showing in these results do not arise as a result of risk aversion on the part of decision makers. A risk-neutral investor would also have a value to waiting, since by waiting they may be able to raise the expected value of the project. In this sense, any increased threshold arising from the option value of waiting is a result of perfectly rational economic behaviour. For this reason, we avoid inferring that uncertainty creates a "barrier" to investment as this could imply that the behaviour is in some way sub-optimal and requires direct policy intervention. Policy intervention to reduce the uncertainty itself may indeed be warranted, but it should not be inferred that a distinct policy is required to overcome an investment "barrier". Risk-averse investors would tend to have an even greater incentive to wait in order to avoid downside risk, so our assumption of risk-neutrality would tend to underestimate the value of waiting.

In addition, the value of information to be gained by waiting will depend on the quality of this information. In our model, we have assumed that perfect information about the subsequent price of carbon is delivered immediately in the year of the jump. In reality, a policy information event is unlikely to be so orderly, with information being gained gradually in the years leading up to the event, and price discovery taking some time to be established in the years after the event. This assumption of very high "quality" of information would tend to overestimate the value of waiting. Our assumption of prices jumping with equal probability anywhere in the range ± 100% may also be an overstatement of the actual uncertainty facing companies. It nevertheless provides a relatively simple starting point to launch the discussion.

Modelling prices

Each element of the cash flow is potentially uncertain and introduces an element of risk into the project. However, the exposure of any given investment to these risks will depend on the specific nature of the electricity market for which it is being considered. For example, the price risks in a coal-dominated market will be different from the price risks in a gas-, nuclear- or hydro-dominated market. The situation is also complicated by the fact that several of these parameters are interrelated due to the way in which different prices are formed in electricity markets. The following sections describe how these relationships are represented in the model. In Chapter 4, we further discuss the importance of different types of market in the real world.

Relationship between electricity price and fuel price

Price formation for electricity will vary according to market structure, but should reflect the operating costs for the generating plant in the system. The level of demand at any given time in the electricity system as a whole cannot exceed the total available capacity in the system, and is usually significantly less. Not all power plants will be needed to meet this demand, so there has to be a priority order (called "merit order") in which they are switched on. Usually this will be in order of cost - specifically, those plants with the lowest short-run marginal cost (including fuel and operating costs, but not capital cost repayments) will be first in the merit order, and those with higher short-run marginal costs will be further up the merit order. The merit order typically comprises renewable sources (including hydro) and nuclear power at the bottom end, as these tend to have low operating and fuel costs. The next plant in the merit order is usually a base-load plant, typically coal fired and/or gas fired. At the top of the merit order is a plant with higher variable operating costs (typically an older or less efficient plant), which may run on coal, oil or open-cycle gas turbines.

The price of electricity will normally be driven by the short-run marginal cost of the last plant in the merit order to be dispatched (the marginal plant). Since the plants at the top of the merit order tend to be fossil-fuel based plants, the electricity price is usually strongly driven by fuel prices – the particular relationship depends on which type of fuel based plant tends to be the marginal plant on average over the year. An illustrative example is shown in Figure 8.

FIGURE 8

A schematic generation stack showing the merit order of dispatch in a competitively priced electricity system

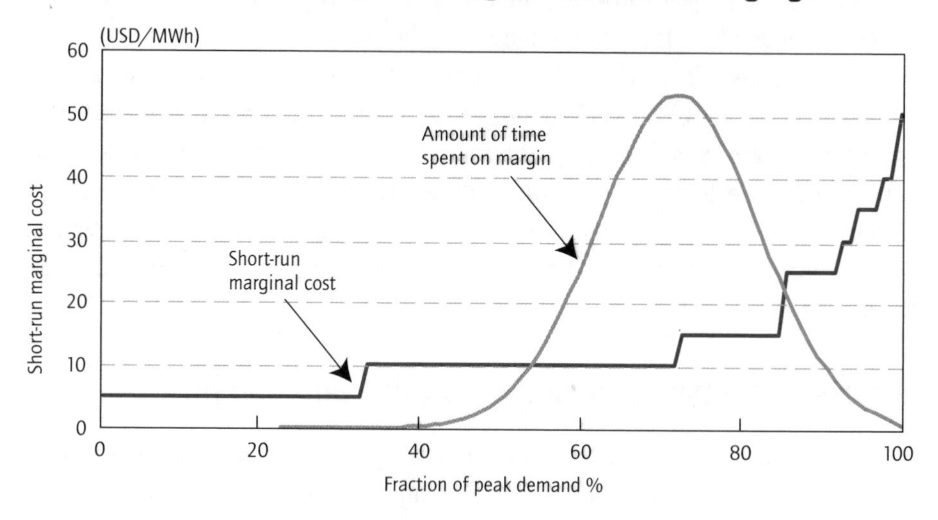

A plant should normally be dispatched in order of increasing short-run marginal cost until there is sufficient supply to meet demand. The last plant to be dispatched is the marginal plant, setting prices for that particular demand level. Because demand levels vary during the year, the type of plant that acts as marginal plant also varies.

Expectations of the price formation process over the lifetime of the plant are a very important determinant of risk. For example, if an investor could be certain that a gas plant will always be on the margin, then investment in a gas-fired plant would be low risk, even in the face of relatively high gas price volatility, because as long as the new plant can "beat" the competing gas plants in the merit order, it will be dispatched and show a profit. Alternatively, if an investor could be certain that a coal-plant will be on the margin and therefore setting electricity prices, then a gas plant might be considered more risky since there would be greater exposure to the gas prices if electricity prices are not correlated with the gas price. In a more realistic scenario, where the marginal plant is not known, because it can vary according to fuel and CO_2 prices, then both types of plants would show some degree of fuel price risk – the degree would depend on the probability of a change in the merit order.

In the model, we take various different cases for the price formation process, including the following assumptions:

- the electricity is a "free" variable, unrelated to fuel prices;
- variations in electricity price are closely correlated with gas price; and
- electricity prices are driven entirely by the marginal plant, and that this could be either gas, coal or vary between coal and gas depending on underlying CO_2 and fuel prices.

Relationship between CO_2 prices and fuel prices

In this report, we use CO_2 price to represent the costs to companies of meeting a given environmental constraint through any type of regulatory instrument. The way in which this CO_2 "price" is experienced by a company will depend strongly on the type of policy instrument used. A CO_2 tax, for example, would simply determine the CO_2 price directly, whereas imposing a certain technology standard would impose an effective price that would relate to technology costs. In these two cases, there would not be a direct link between the CO_2 price experienced by the company and other underlying quantities such as fuel prices.

Under an emissions trading scheme, in contrast, the CO_2 price is (at least in theory) determined by the marginal cost of abatement of CO_2. In the European Union (EU) Emission Trading Scheme, for example, a large proportion of the emissions in the scheme come from the power sector, and a key short-term abatement option in this sector is an operational switch from coal-fired generation to gas-fired generation, which has lower emissions per unit of electricity generated. In order for a gas-fired power plant to be dispatched instead of a coal-fired plant, the short-run marginal costs for the gas plant would need to be lower than for the coal plant. This will occur if the CO_2 price is sufficiently high, since coal plants emit more CO_2. The threshold CO_2 price required to drive this operational switch will also depend on the prices of gas and coal, since these largely determine the gap in short-run marginal costs between the two technologies that must be filled by the CO_2 costs. Using some simple assumptions about the efficiency of the gas and coal plant in the electricity system, it is straightforward to calculate the threshold price at which a switch from coal-fired plant to gas-fired plant would be expected to occur (Figure 9).

FIGURE 9

Price of carbon at which coal and gas would change places in the merit order

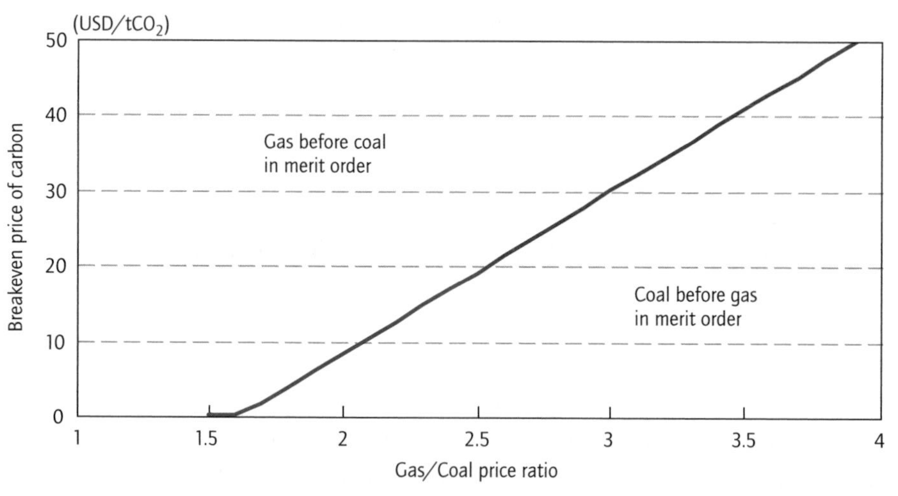

The price of carbon at which the short-run marginal cost for a gas plant equals the short-run marginal cost for a coal plant.[6] If switching from coal to gas in the dispatch order of power stations is expected to determine the marginal price of CO_2 in an emissions trading scheme, then CO_2 prices will vary in line with gas price fluctuations according to this type of relationship.

The implication of this is that there is a functional relationship between the price of gas and the price of CO_2 in an emissions trading scheme, so that carbon and gas price fluctuations would be expected to be well correlated. Experience to date in the EU Emission Trading Scheme (ETS) is that this relationship has held over large ranges, and over long periods, despite some disruptions to the CO_2 price from a variety of sources. Prices of CO_2 have therefore been highly correlated in the EU ETS, so for some of the modelling runs described in the next section, we take a high degree of correlation between CO_2 and gas prices to represent this case.

If this relationship were assumed to hold rigorously, then CO_2 prices would always find a level such that gas came before coal in the merit order. However, in the EU ETS, the coverage of the scheme is wider than the coverage of any individual electricity market since it includes all 25 EU member countries, and allows credits

6. In calculating this relationship, the gas plant efficiency is assumed to be a factor of 1.5 times greater than the coal plant efficiency. Variable operating and maintenance costs of USD 3.33/MWh and USD 1.5/MWh were assumed for coal and gas plants respectively. The position and angle of this breakeven line would be different if these assumptions were changed.

from joint implementation and clean development mechanism projects from non-EU countries to be used. Although the relationship with gas prices may retain some influence on CO_2 prices (*e.g.* through a correlation between price fluctuations for the two commodities), we do not assume in general, that the CO_2 price necessarily is sufficient to force coal higher than gas in the merit order. The assumption about what plant is higher up the merit order is allowed to be taken independently in the model.

There is also a causal link in the other direction. Wholesale gas prices could be affected by CO_2 prices if they drive demand for gas up as a replacement for coal. We do not explicitly include this effect in the model, since gas prices are exogenously specified according to the process described in Figure 5. We do not assume that a jump in CO_2 price due to policy intervention would have a knock-on effect on gas prices. Further discussion of this point is given in Chapter 4.

Pass-through of CO_2 price to electricity price

The assumption that electricity prices are driven by short-run marginal costs leads to the incorporation of CO_2 price into the electricity price, either because each tonne of CO_2 emitted must be paid for directly (*e.g.* in the case of a tax or auctioned tradable permits), or because emitting the CO_2 results in an opportunity cost relating to the permits that would otherwise have a saleable value on the market (in the case of "grandfathered" permits).[7] In the case of other (non-market based) policies, the "pass-through" of environmental costs to electricity prices may be less direct, and will depend on the way in which companies are able to recover the capital costs of their investments. In emissions trading schemes, the extent to which CO_2 costs might be passed through to the electricity price are disputed, and may depend on the level of free allocation to existing and new plants and the ownership structure of the market. (See, for example, Baron and Reinaud, 2006; CPB, 2003; ECN, 2005; Ilex, 2004; Shuttleworth *et al.*, 2005.)

In general, in this report, we assume that the CO_2 price is passed directly through to the electricity price at a rate determined by the emission rate of the marginal plant in the merit order. These pass-through rates depend on the type of plant at

7. A provision in a statute that exempts those already involved in a regulated activity or business from the new regulations established by the statute.

the margin of the merit order, which in turn depend on the type of market - the rates will be different depending on whether the markets are dominated by coal, gas, nuclear or hydro.

In a system with both coal and gas generating plants, the merit order itself can change as a result of CO_2 and fuel price changes, so the pass-through rate is also another source of uncertainty. Figure 10 shows how the electricity price depends on the CO_2 price and how the slope of the curve depends on which plant type is on the margin of the merit order.

FIGURE 10

When coal and gas switch places in the merit order, CO_2 price will feed through to electricity price at a different rate

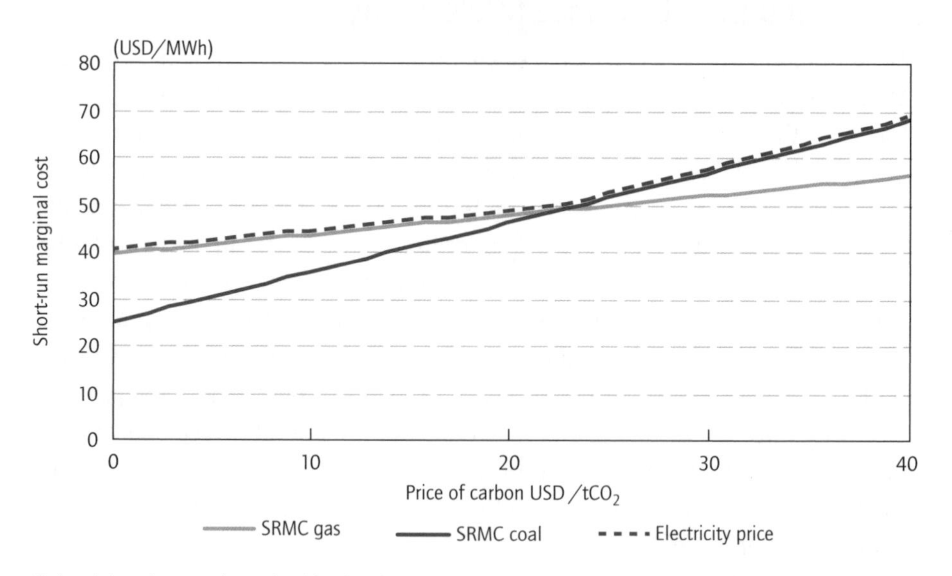

If electricity prices are determined by the short-run marginal cost of generation, there will be a relationship between electricity prices and CO_2 prices similar to that shown in Figure 10. At low carbon prices, a gas plant will be on the margin setting electricity prices. At high CO_2 prices, a coal plant will be on the margin setting electricity prices. The crossover point is determined by fuel price and plant efficiency as described in Figure 9.[8]

8. In this figure, the gas and coal fuel price ratio is taken to be 2.7.

As noted earlier, expectations of the electricity price formation process (including CO_2 pass-through) will be an important determinant of risk for a generation project. Figure 10 suggests that the rate of CO_2 price pass-through itself may be another source of uncertainty, although how important this is will depend on what the expectation is when the financial case is made for the project. Under current price conditions in most IEA countries with a mixed portfolio, gas would be higher up the merit order than coal. If a project financial case is made with the assumption of gas on the margin determining the pass-through rate, then any deviation from this towards coal setting the pass-through rate would increase the electricity prices above expectations, leading to higher than expected returns on the investment. Conversely, if the project is justified on the basis of the higher feed-through rate set by coal, and conditions change such that a lower feed-through rate applies in practice, then the project would under perform compared to expectations.

Limits of the modelling approach

The above discussion considers prices for electricity, fuels and CO_2 to be external to the investment decision (*i.e.* that investors are price takers). This is also the assumption made in the quantitative analysis presented in Chapter 3, where it is assumed that the investment decision itself does not affect price expectations (or at least that the price expectation used in the analysis includes the effects of the investment on the market). However, this assumption ignores the possibility that other investors, who experience the same price signals, will make the same investment decision, and that the market may be driven towards unpredicted outcomes where prices are affected by collective decisions at the sector level. Such feedbacks and strategic aspects of decision making are discussed in more detail in Chapter 4.

INVESTMENT THRESHOLDS

This chapter presents the quantitative results of the modelling using the approach laid out in Chapter 2. In summary, the model calculates an investment threshold as an additional risk premium for various power generation investment projects facing uncertain cash flows. The uncertainties modelled include fuel price risk and CO_2 price risks, which are used as a proxy to represent policy uncertainty. The relationship between CO_2, fuel and electricity prices are generally reflective of an emissions-trading type policy mechanism operating in the context of a competitively priced electricity market. Some different assumptions are also made to look at the effects of different types of policy and different types of electricity markets.

The analysis does not attempt to represent all the risks faced by companies making power generation investments. The assumptions have been kept simple to keep the analysis as transparent as possible. The aim is to provide an illustration of the importance of climate policy risk in relation to other key risks and to make the analysis as broadly relevant to IEA member countries as possible.

CO_2 price risk

We begin the analysis with some examples where we make CO_2 prices uncertain and keep the fuel prices deterministic. This helps to build understanding about the effects of CO_2 price uncertainty before going on to the more complicated case where we also include fuel price uncertainty.

Choice between new coal and new gas plants

We start the results section with a relatively simple example of the choice between investing in a new coal and a new gas generation plant in the case where only CO_2 prices are uncertain, and all other prices and costs are known. This is chosen as a representation of the basic base-load generation investment choice facing many companies and sets the scene for the more complex cases presented in the following sections.

In these runs, CO_2 prices are stochastic, with an annual random walk variation of ±7.75%, and a price jump occurring in year 11, allowing a period of 10 years of relative stability before the jump. The expected (mean) CO_2 price is USD 31/tCO2, set at a level that equalises the financial case for a new coal and a new gas plant under the technical and fuel price assumptions used.

Gas and coal prices are taken to be fixed for the duration of the project's life at USD 5.6/GJ and USD 1.9/GJ respectively. It is assumed here that electricity prices are determined by the short-run marginal costs of a gas plant. This means that electricity prices are deterministic except for the inclusion of CO_2 costs, which feed through to electricity prices at a rate determined by emissions levels of a 40% efficient gas plant (variations on this assumption are made in later sections). With this pass-through rate, the coal plant does worse from increases in CO_2 price, whereas a new efficient combined cycle gas turbine (CCGT) plant will do better from an increase in CO_2 price.

Figure 11 shows the investment threshold expressed in terms of the additional gross margin required to overcome the value of waiting for coal and gas plants. The x-axis is the year in which the investment decision is made, and the y-axis is the additional expected return on investment (over and above a zero NPV) required to stimulate investment in that year. The less time there is available before the jump in prices in year 11, the greater the expected return on investment would be required for the decision to be taken (*i.e.* the threshold that needs to be exceeded if the investment decision is to be made in year 6 is considerably greater than the threshold in year 1). This follows the discussion developed in Chapter 2. The y-axis shows the threshold expressed in terms of the additional return required per kilowatt (kW) of capacity of the generation plant being considered. This can be compared with capital costs for the plant of approximately USD 1300/kW for coal and USD 600/kW for gas.

These results show an interesting technology interaction effect. Two sets of results are plotted together in Figure 11, one with the options considered separately (single options) and the other with the two options as alternatives (dual options). The breaks in the curve occur when the model could not determine the investment threshold close to the CO_2 price jump. Because the gross margin of the two options respond in opposite directions to a carbon price jump, they act as a good hedge for each other, and the value of waiting to see which direction carbon prices jump becomes significantly higher than if they were considered on their own.

This technology interaction effect can be appreciated logically. Consider the case where the gas plant is considered separately - the option is to build a plant and

FIGURE 11

Investment thresholds for the coal versus gas investment decision under CO2 price uncertainty

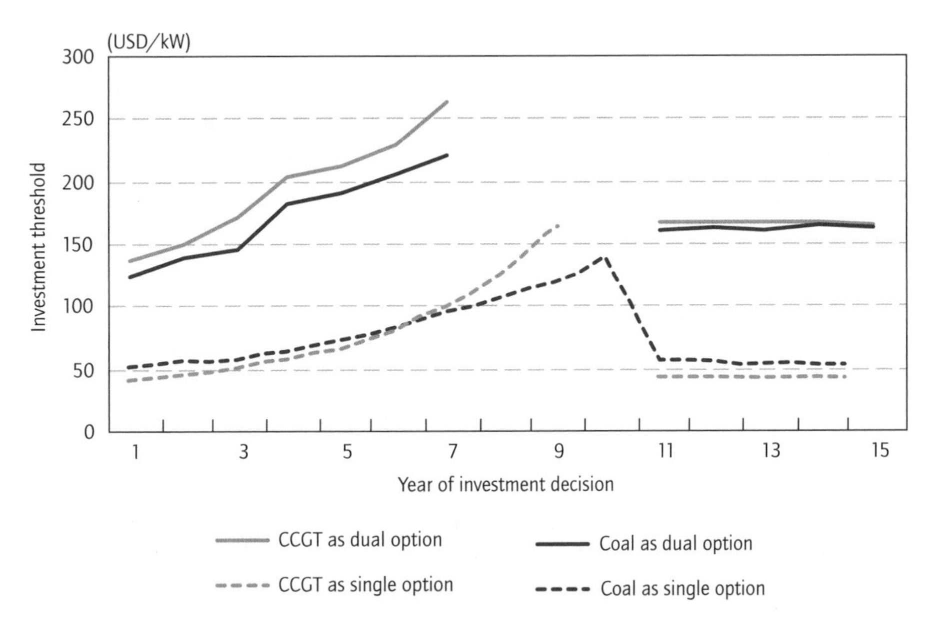

The CO_2 price uncertainty includes a price jump in year 11. The expected gross margins have to exceed capital costs by this amount in order for investment to proceed. The investment threshold gets bigger the closer to the time of the price jump the investment decision is made.

generate income, or not to build a plant and generate no income. If the build decision were postponed until after the carbon price jump, then if carbon prices go up, the plant would be built, and if prices go down, then no plant is built. The expected value of waiting would therefore be calculated from the average of a zero cash flow case (where the carbon price goes down making the plant non-viable financially) and an increased cash flow case (where carbon prices go up improving the CCGT plant gross margin).

Now consider the case where the option is either to build CCGT plant or to build a coal plant. The company will build either a coal plant in the case where carbon prices go down or a gas plant in the case where carbon prices go up. This value of waiting is significantly higher than the value of waiting for a single option because there is no zero cash flow outcome – the expected value of the optimal choice is always positive.

The implication of the technology interaction effect is that the investment threshold required to incentivise immediate investment in the gas plant depends not only on the value of waiting for the gas plant project itself, but by the value of waiting for other alternative investments that are available concurrently.

Because CO_2 price is the only stochastic variable, we can show the investment threshold in terms of CO_2 price, which further helps to visualise the impact of considering these technologies together as alternative investments. Figure 12a translates the investment threshold given into a threshold carbon price. The results show that when gas and coal are considered side-by-side as alternative investments, the effect of carbon price uncertainty can be to open up quite a considerable range of carbon prices where the optimal response would be to wait. The "wait" zone is larger when the investment decision is made close to the year in which the carbon price jump occurs (year 11). The break in the two lines in years 8–10 arises from a limitation of the model: in principle a threshold does exist for those years, but the model is unable to provide a value for the investment threshold this close to the jump.

The gross margin of the project also responds to electricity prices. Therefore, if expectations of future electricity prices rise sufficiently, then the project value can become high enough to overcome the value of waiting. We can therefore determine the increase in expected electricity price required to exceed the investment threshold. The results can be interpreted as follows: for any given year in which the investment decision is being made, in order to close the "wait" zone shown in Figure 12a, expectations of future electricity prices would need to be increased by the amount shown in Figure 12b. We can estimate from Figure 12b that in order to incentivise immediate investment in either coal or gas, the electricity price would have to rise by about 5-6% if the investment decision is made 10 years before the price jump. It would have to rise around 8-9% if the investment decision is made 5 years before the price jump. These are not large rises – electricity prices can move by this order of magnitude in response to a single large investment.

Carbon price uncertainty may therefore lead either to a delay in investment or to a modest increase in electricity prices to meet the threshold, justifying immediate investment. Which of these effects is manifested may depend on the ownership structure of the power market – monopolies may behave differently in this regard to oligopolies or competitive markets. In any case, the two effects are linked since in the context of increasing demand, a delay in investment in the market will lead to higher electricity prices. These factors are considered in more detail in Chapter 4.

Figure 12

Carbon price and electricity price thresholds

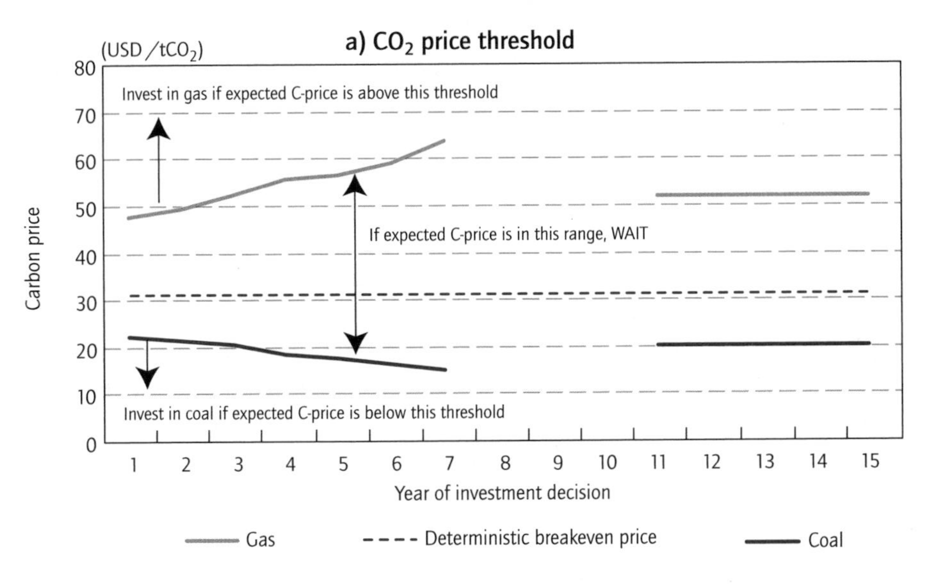

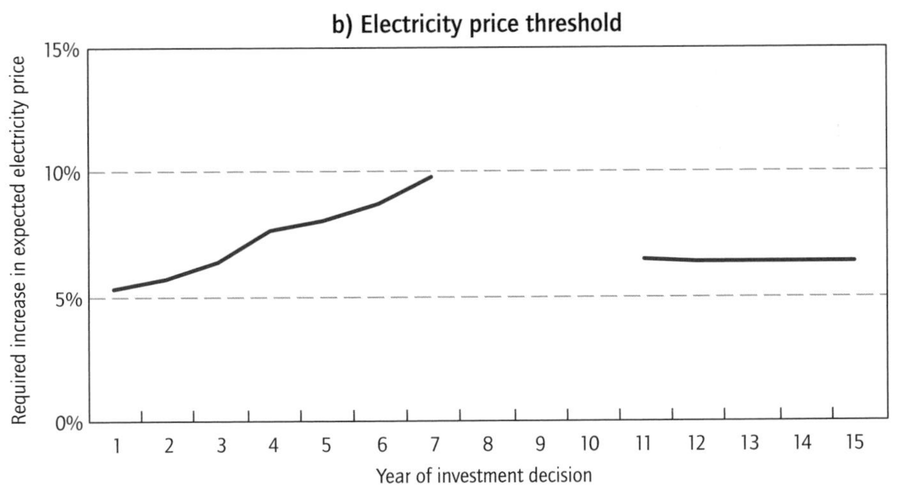

Figure 12a: The investment threshold is expressed in terms of carbon price. If expected future carbon prices are above the top line, then immediate investment could occur in gas, and if they are below the bottom line then immediate investment could occur in coal. In between, no investment would occur – the option to wait would be exercised. This is evaluated at an electricity price that exactly meets the threshold for investment under deterministic conditions as calculated by a discounted cash flow.

Figure 12b: Alternatively, the threshold can be expressed in terms of the increase in expected electricity price required to overcome the value of waiting and incentivise immediate investment in the presence of carbon price uncertainty. This electricity price surcharge to the end-users is between 5% and 10%. If expected electricity prices were raised by this amount, this would close the "wait" zone in Figure 12a.

Replacing an existing coal plant with a new gas plant

In this example, we look at the investment case for a new gas-fired plant as a replacement for an existing coal-fired plant. The coal plant is assumed to have a useful remaining technical life of 25 years, and to be just about profitable under the expected (mean) conditions, but the replacement gas plant would be significantly more profitable under expected conditions and would justify immediate replacement under a normal DCF analysis. It is assumed that both plants would operate as base load with 85% utilisation rates.

However, this conclusion changes when future prices are uncertain. Since the capital costs of the coal plant are assumed to be sunk, it can operate at very low levels of the gross margin and remain profitable. Uncertain future carbon prices will cause a variation in the gross margin, but losses are capped at the fixed operating costs (USD 30 000/MW/yr), since at this point the plant could simply be switched off. Figure 13 shows the distribution of a coal plant gross margin under carbon price uncertainty, where the price jump due to policy uncertainty occurs in year 11. Note that because of the curtailment of the downside risk for the existing plant, the expected (mean) gross margin increases after year 11.

FIGURE 13

Distribution of annual gross margin for coal power under CO_2 price uncertainty

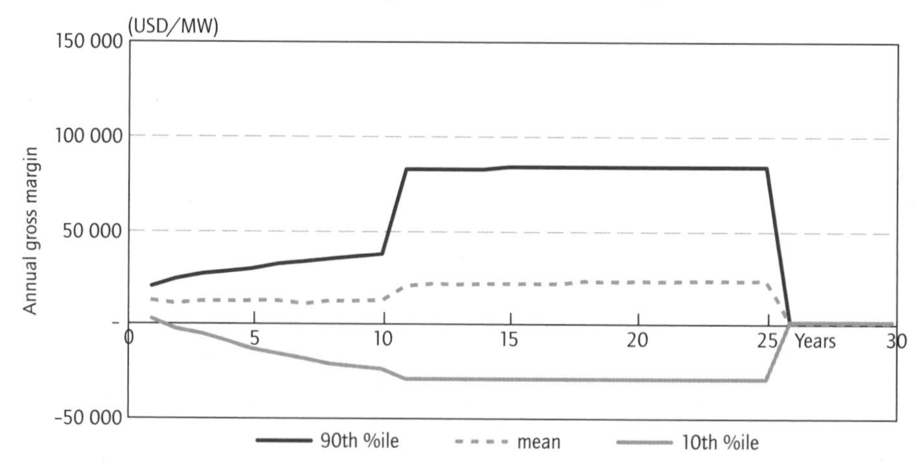

The distribution of the annual gross margin for an existing coal plant under CO_2 price uncertainty was calculated with an assumption of a carbon price jump in year 11 representing policy uncertainty. The downside losses are capped at the level of fixed operating costs.

When considering replacement of this existing plant with a new gas-fired plant, the relevant cash flow to be considered is not the stand-alone gross margin for the new plant, but the relative difference in the gross margin between the existing plant and the new plant.

The asymmetry in the distribution of future cash flows for the coal plant therefore also affects the investment case for the new gas plant. This distribution is shown in Figure 14. Since the downside for the coal plant is capped, this leads to a capping of the upside in the relative gross margin for the gas plant compared to the coal plant. The expected relative gross margin for the plant replacement project therefore drops after year 11.

FIGURE 14

Distribution of the relative gross margin for a new gas plant

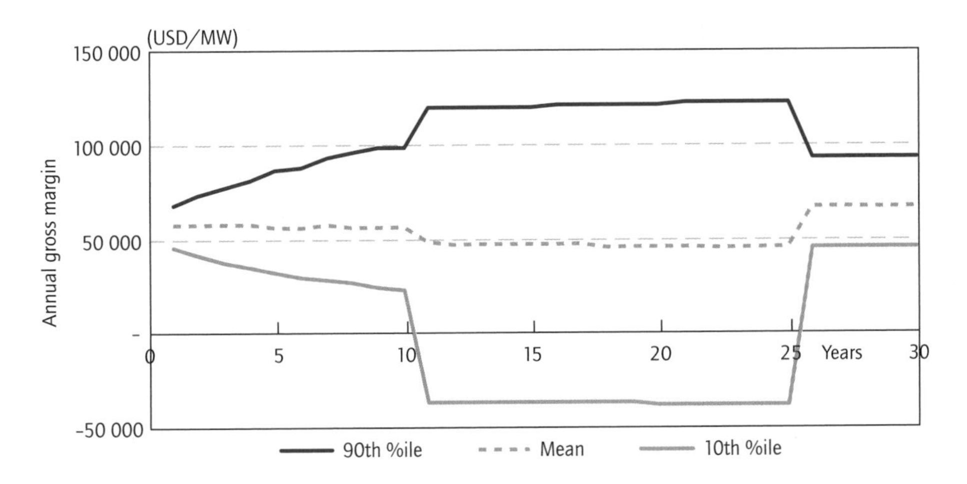

The upside gains are limited by the capping of losses in the coal plant losses. The drop in the expected (mean) future gross margin from the gas plant relative to the existing coal plant means that the investment threshold for the new gas plant is higher than in the case where the gas plant investment is considered as a stand-alone project. This can be seen from the higher threshold values as shown in Figure 15 compared to Figure 11. This is because policy uncertainty creates an option value to maintaining an existing plant, even when there are apparently more profitable options available. The threshold drops off in later years because the residual option value of maintaining the less profitable existing coal plant diminishes towards the end of its life.

The main conclusion from this is that the greater the level of uncertainty, the more likely it is that companies will aim to keep low capital cost options open such as maintaining and possibly extending the life of an existing plant.

FIGURE 15

Investment threshold for replacing an existing coal plant with a new gas plant under CO_2 price uncertainty

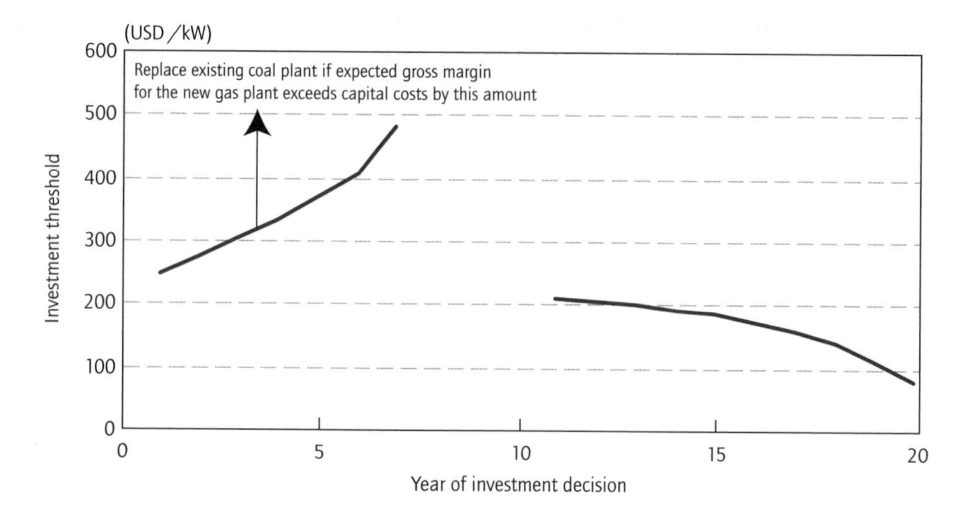

Carbon capture and storage

In this section, we extend the previous analysis to allow for subsequent retrofit of carbon capture and storage to gas and coal plants. The analysis here still only deals with CO_2 price uncertainty, ignoring fuel price uncertainty. This is not a serious omission, since the retrofit of carbon capture and storage is almost entirely driven by carbon price. It is not very sensitive to fuel price variations.

The technology for carbon capture and storage is described extensively in the literature (see for example IPCC, 2005; IEA, 2004b). It essentially involves removing CO_2 from the combustion process, transporting it in a pipeline, then pumping it deep underground into a suitable geological storage site. We can avoid discussing the technical detail here, and simply view CCS as an investment option that significantly reduces emissions of CO_2 in return for a capital investment, a reduction in efficiency of power generation (accompanied by a loss in revenue due to additional on-site power demand) and an increase in fixed and variable operating costs to cover the cost of capture, transportation and storage. (Cost and performance assumptions are listed in Appendix 1.)

The extent of abatement of carbon emissions from a power plant depends on the specific technology used and the site-specific details of the storage site. In this study, we have followed the IPCC figure of an 86% reduction in emissions resulting from retrofitting the existing technology (IPCC, 2005). We assumed that this reduction is known; although in reality, there will be some technical risk and uncertainty associated with the actual abatement level achieved and the possibility of future leakage of CO_2 from the storage sites. Technical risks are discussed qualitatively in Chapter 4, but a more detailed assessment of the interaction between climate policy risks and technical risks for CCS is left for future work.

The plant continues to be fired with the same fuel, so apart from the loss in efficiency, the economics of carbon capture and storage as a retrofit option is not strongly dependent on fuel price, but much more strongly dependent on carbon price. The option to build a CCS plant is modelled as a modular investment – first the investment decision to build the power plant (either coal or CCGT) has to be taken, and then the retrofit of CCS as a further option is modelled. We assume that there is not a major cost penalty incurred by investing subsequently as a retrofit rather than in a single investment. The coal and gas plant are assumed to be fully "capture ready".

As before, we are aiming to evaluate the effects of uncertainty at the point where CCS would normally be considered cost-effective. Figure 16 maps the break-even points for the different investment options under price certainty using a DCF analysis. The red lines show the values of CO_2 and fuel price at which the net present value (gross margin minus capital costs) of the different technologies shown would be equal. The actual positions of these lines depend on assumptions made about technology costs and different assumptions, which would lead to different break-even points. For our analysis, the actual position of these lines is not particularly important, except insofar as they provide a reference point for us to measure the impact of introducing uncertainty.

FIGURE 16

Breakeven points for carbon capture and storage technology

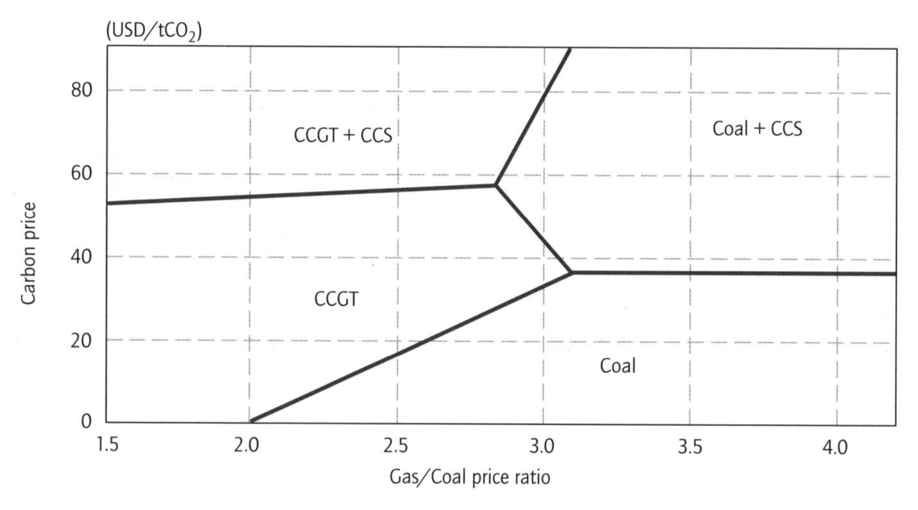

The blue lines show where the net present value for any two technologies would be equal under deterministic prices. The axes represent expected prices over the duration of the project for carbon and the ratio of gas/coal price.

If prices are uncertain, then as we have seen before, there is an option value of waiting, which has to be overcome in order to incentivise immediate investment. In Figure 17, we represent the effect of introducing CO_2 price uncertainty, using the standard case of 7.75% annual volatility (geometric Brownian motion) and a price jump in the range ±100%. Figure 17a shows the results if the jump occurs in year 11 and Figure 17b shows the results if the jump occurs in year 6. The bold black lines represent results of the model expressed in terms of the change in price (carbon price or gas/coal price ratio) required to overcome the investment threshold given by the model in year 1. The shaded areas are extrapolations of the actual model results to give a postulated region where the option to wait might be exercised.

FIGURE 17

Impact of CO_2 price uncertainty on CCS

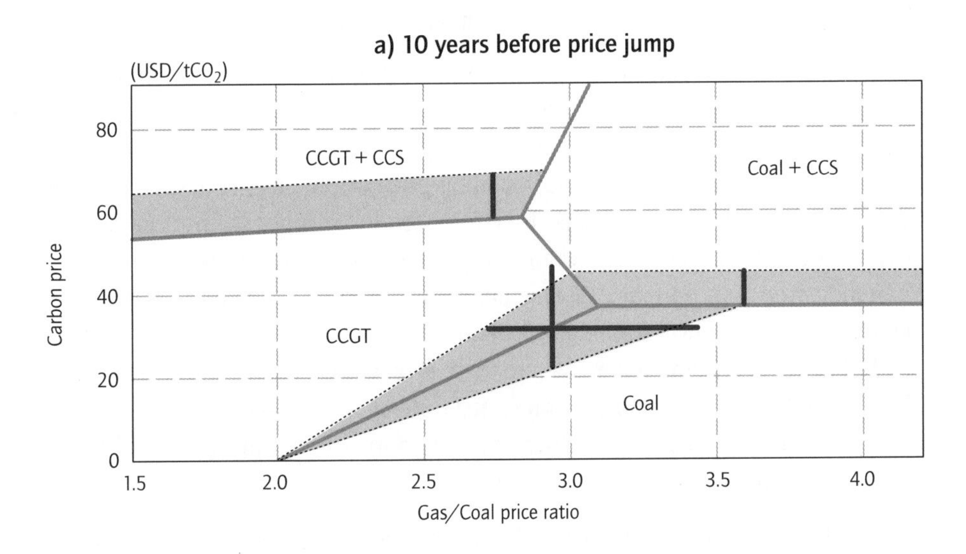

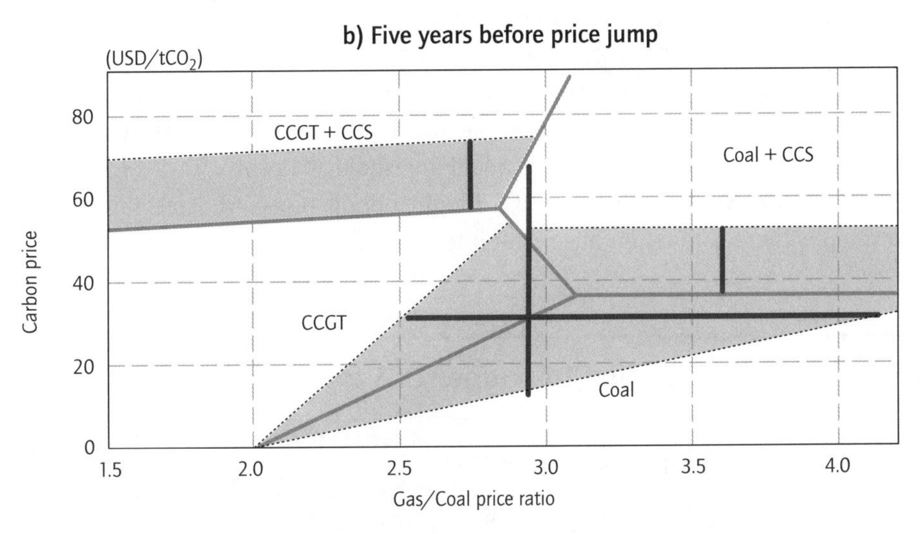

The bold black lines show the change in prices required to overcome the investment threshold. The shading is an extrapolation of these model results to indicate the region in which the option to wait is likely to be exercised. This "waiting" region is significantly larger in Figure 17b where there is only 5 years before the price jump, compared to Figure 17a when there is a 10-year period before the price jump.

The effect of carbon price uncertainty on the choice between CCGT and coal indicated in Figure 17 is exactly analogous to the results in Figure 12a. In that discussion, we noted that the option to wait would not be exercised if the electricity prices rose by a modest amount (5-9%), sufficient to overcome the value of waiting.

However, electricity and fuel prices are not strong drivers of the investment case for the retrofit of CCS technology. So unlike the case for the choice between CCGT and coal plants, an increase in electricity price would not tend to close this "waiting" region - only an increase in carbon price would be able to overcome the value of waiting. These results therefore tend to suggest that immediate investment in CCS technology would only occur if the carbon price were higher than indicated by a normal DCF calculation. In the case where there is a 10-year period of relative price stability, the required increase in carbon price is quite modest (~15%), but in the case where there is only 5 years before the price jump, the required increase is more significant (~35%). The increase in carbon price shown in Figure 17 required to drive immediate investment in the face of future price uncertainty is reproduced in Table 2.

The existence of an option to retrofit CCS technology to a coal plant has an interesting effect on the investment case for the coal plant itself. One of the key risks facing investment in coal is the possibility of carbon prices being higher than expected. The possibility of retrofitting to CCS technology later acts as a good hedge against this risk, and reduces the investment risk for coal. This is shown in Figure 18 by comparing the investment threshold for coal with a CCS retrofit option (Case B) and without a CCS retrofit option (Case A). Note that in this example, there is no alternative option to invest in gas.

TABLE 2

CO_2 price to trigger CCS investment

	Expected CO_2 price required under certainty (DCF calculation) USD /tCO_2	Expected price required with uncertain CO_2 price jump in year 11 USD /tCO_2	Expected price required with uncertain CO_2 price jump in year 6 USD /tCO_2
CCS retrofit to coal	38	44	52
CCS retrofit to CCGT	57	67	77

Note: The numbers show the expected CO_2 prices required in order to exceed thresholds for investment in a CCS plant.

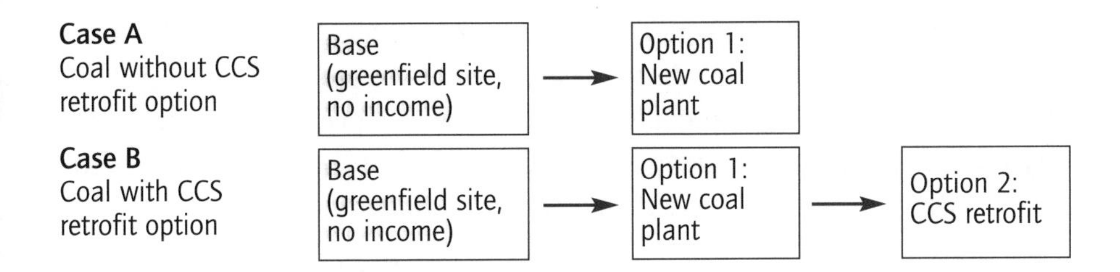

Figure 18

CCS acts as a hedge against uncertain CO_2 prices, accelerating investment in a coal plant

Case A: investment rates for coal without CCS retrofit option under CO_2 price uncertainty, with a jump in year 6

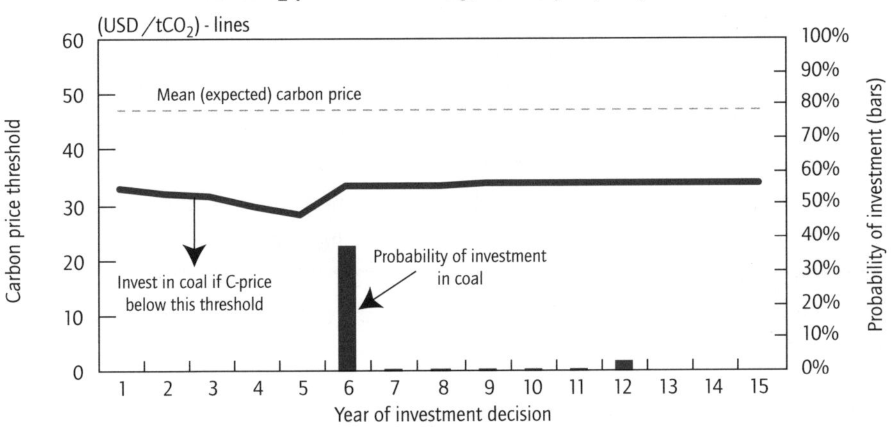

Case B: investment rates in coal with the option of future retrofit of CCS under CO_2 price uncertainty, price jump in year 6

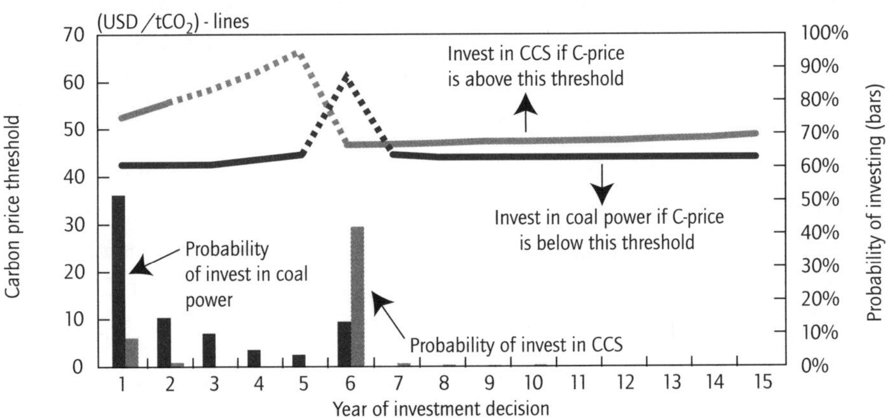

For both Case A and B, the expected mean carbon price is USD 48/tCO_2 with a price jump expected in year 6. At this average price, the probability of exercising the option to invest in a coal plant without future possibility of a CCS retrofit is very low. In Case A, a coal plant is only built after year 6 if carbon price drops sufficiently. The total probability of building a coal plant is 45%.

With the option of adding CCS technology at a later date (Case B), the CO_2 price threshold below which investment in coal would be justified is raised from around USD 32/tCO_2 in Case A to around USD 42/tCO_2 in Case B, bringing it closer to the expected value for the run. This increases the probability of investing in year 1 from zero to 49%. The total probability of investing in coal over all years in Case B is 100%. Only about 50% of the time is a plant with CCS technology itself actually built, mostly after the price jump in year 6 if prices go up.[9]

In the case of a CCS technology retrofit to a CCGT plant, the situation is less clear, since the NPV for both technologies increases as carbon price increases, so a plant with CCS technology does not have such a clear hedging value. However, this conclusion relies on our original assumption that the electricity price is set by a gas plant being the marginal technology in the merit order - this assumption is very likely to break down if carbon prices were in the region of USD 60/tCO_2 and above required to make CCS technology cost-effective on a CCGT plant. The boundary between a CCGT plant and CCGT plant with CCS technology shown in Figure 17 is calculated assuming the CCGT plant is already in place.

The method used to represent the investment thresholds for coal, gas and carbon capture and storage in provides an intriguing way of thinking about investment conditions for an integrated gasification combined cycle (IGCC) plant versus a pulverised fuel (PF) coal plant. An IGCC plant essentially uses a two-step process - first coal is gasified to create a gaseous fuel, and then this gaseous fuel is burned in a combined cycle plant very similar to (and to some extent interchangeable with) a CCGT plant used for a natural gas fired plant.

Assumptions about technology costs and efficiency, will affect the estimated paybacks of these two technologies. For example, an IGCC plant for full-scale power generation is a less proven technology than a PF coal plant; therefore, the technical risks are higher. However, to a reasonable approximation an IGCC plant with CCS technology as a whole investment opportunity would be expected to

9. Solid lines are actual model results; dashed lines are extrapolations.

have a similar profile in terms of exposure to fuel price and CO_2 price risk as a PF coal plant plus CCS technology as a single investment.

For an IGCC plant plus CCS technology, we can consider the investment as a three-step process:

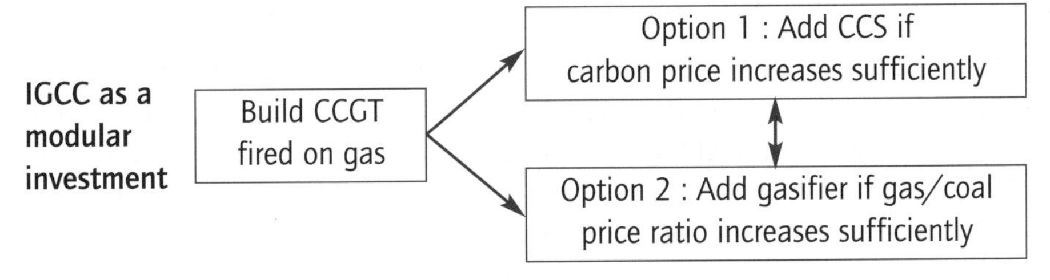

How the investment proceeds from a CCGT plant to an IGCC plus CCS plant would depend on the particular price path, but one could envisage that starting from a CCGT plant, CCS technology would be added if the carbon price increases sufficiently and then a gasifier would subsequently be added if the gas/coal price ratio rose sufficiently. These different development paths are indicated schematically in Figure 19. So in addition to technology considerations, the choice between an IGCC and PF coal plant may also depend on initial expectations about the price paths for gas and carbon prices.

Given the expected similar financial performance of an IGCC and PF coal plant, we would expect the effect of carbon price uncertainty to be similar for the two.

Figure 19

Possible development paths for IGCC and pulverised coal

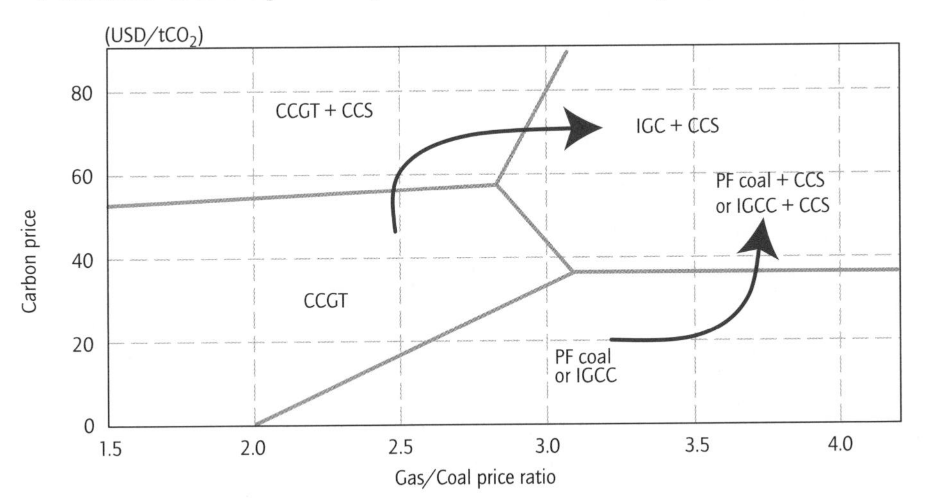

However, does the flexibility of an IGCC plant to be able to fire on either coal or gas give it an advantage over a PF coal plant in terms of project value? We can test this by looking at the value of the project under conditions of varying price with our previous assumptions (*i.e.* with 7.75% annual volatility with a geometric Brownian motion price process).

Once the capital cost of the gasifier had been sunk, the gas price would have to drop to below half its expected value to justify a switch back to gas (*i.e.* the gas/coal price ratio would have to drop from around 3 to around 1.5 or below). Under the fuel price assumptions we used in the model, this does not happen very often, but this operational flexibility would add around 9% to the expected net present value of the project compared to a normal deterministic analysis.

The ability to "switch off" the carbon capture and storage plant under these assumptions appears to be quite valuable. Because CCS technology increases the operating costs and reduces the efficiency of the plant, there is an economic incentive to switch it off if the carbon price drops below about USD 21/tCO_2 under the cost assumptions used in the model. Again, this is a large drop in carbon price (compared to the expected value of around USD 38/tCO_2 used in modelling CCS retrofit to coal). Nevertheless, under our assumptions of a price jump in the range ±100%, this condition occurs relatively frequently, and the ability to switch off the CCS technology to revert to a normal coal-firing plant would increase the net present value of the project by around 30%. This additional value of operational flexibility would apply both to PF coal plants and to IGCC plants.

CO_2 and fuel price risk

Expanding the analysis

In this Chapter, we extend the analysis by introducing three new factors:

- third generation technology (nuclear power) in the investment choice;
- fuel price uncertainty; and
- different assumptions about the marginal plant in the electricity merit order.

Nuclear power is virtually free of CO_2 emissions and requires almost no fossil fuel input. It is therefore often regarded as being free from the risks associated with uncertainty in gas and CO_2 prices. However, as we shall see in this section, that conclusion does not necessarily hold if electricity prices are determined by the costs of generation by a fossil-fuel fired plant.

We estimate the costs of nuclear power generation using data from the NEA/IEA book, *Projected Costs of Generating Electricity* (IEA, 2005) for this model. Figure 20 shows the levelled cost of generation of nuclear compared to coal and gas using this data. The error bars represent two standard deviations. The average levelled cost for nuclear generation is within USD 1.5/MWh of the average value for a coal plant, although taking into account the range of costs, there is significant overlap in the costs of the three technologies.

FIGURE 20

Comparison of levelled costs for coal, gas and nuclear*

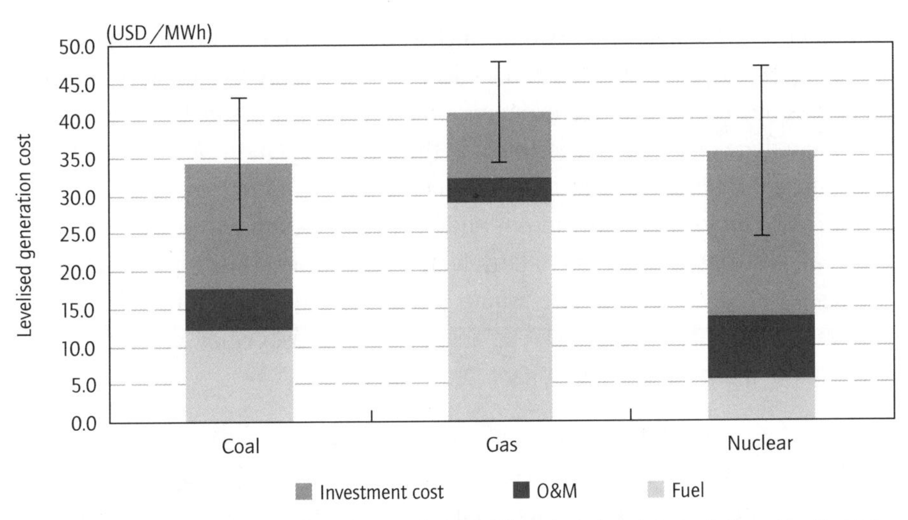

Note: *Levelled costs of generation from the NEA/IEA (2005) book: *Projected Costs of Generating Electricity*. Error bars indicate the two standard deviations range of power generation costs.

Based on these average levelled costs for nuclear, there would be no need for an additional carbon price signal in order to make the technology cost effective. However, the economic case for a nuclear plant would still be sensitive if gas prices were to drop substantially relative to expectations bringing electricity prices with them. We can test the sensitivity of the nuclear investment decision to uncertain gas prices by running the model with gas price uncertainty switched on (7.75% random walk change in annual gas price). The investment threshold introduced would be over 20% of capital costs; the net present value of the nuclear investment would have to exceed USD 264/kWe in order to proceed in the face of fuel price uncertainty. This additional project revenue would be generated if electricity prices were USD 5.5/MWh higher than the levelled cost shown above. This closes the gap of about USD 5.3/MWh between the average levelled generation costs of gas and nuclear shown in Figure 20.

What about carbon price uncertainty? If the average levelled costs shown in Figure 20 were considered a reliable estimate for nuclear costs, then in the absence of other risks, the investment would be profitable even if carbon prices went to zero. Any value of carbon above zero would improve the plant economics, so the investment looks "safe" whatever the value of carbon turns out to be; thus carbon price uncertainty would not lead to any value of waiting.

In addition, the range of costs of a nuclear power plant shown in Figure 20 is quite high, suggesting significant technical and regulatory risk. In most countries, there has been rather limited experience of new nuclear plants being built since the original major round of investment in the technology in the 1970s and 1980s. Companies taking a new decision to build a nuclear plant who want data on plant costs can look at many recent and on-going investments around the world for coal and gas plants, but relatively few for nuclear plants. It is therefore interesting to look at the case where the plant economics for coal, gas and nuclear plants are more finely balanced, and to ask what difference price uncertainty makes to the investment decision.

We therefore set up the model with three investment choices: coal, gas and nuclear. We set the expected prices such that all three technologies have equal (slightly positive) NPVs under deterministic conditions. In order to balance the NPVs for coal and gas plants, we set expected gas prices to USD 52/GJ and carbon prices to USD 25/tCO2. For the case of a nuclear plant, we need to increase costs (capital and O&M) by two standard deviations above the average quoted in the NEA/IEA report in order to make the project NPV balance those for coal and gas plants under these price assumptions. This still puts the cost estimates within the range of some countries' estimates (DTI, 2006). Nevertheless, the fact that we have increased the estimated technology costs should be borne in mind when considering the conclusions drawn from these runs.

In these runs, we also investigate different assumptions about what drives electricity prices. The assumption is made in the model that the electricity price is determined by the short-run marginal cost of the marginal plant in the merit order. Three marginal plant cases are tested:

- a 100% coal plant;
- a 100% gas plant; and
- the model decides whether a coal or gas plant is on the margin in any given year depending on the fuel and carbon prices in that year.

When coal is assumed to be the marginal plant, gas price variations do not pass through to electricity prices, whereas carbon price variations pass through to the electricity price more strongly (since coal is a higher emitter of carbon than gas). When a gas plant is on the margin, gas price variations pass through to electricity prices and carbon price also feeds through, but at a lower rate than in the coal case. In the mixed gas/coal case, the relationship is somewhere in between, although actually closer to the gas case as under the conditions of the run, the model has a greater probability of having gas on the margin rather than coal on the margin.

Earlier results showed that there are two elements to the risk premium created by uncertainty - an element relating to the impact of uncertainty on a single technology, and a technology interaction effect which arises when coal and gas are considered as alternatives rather than as if they are treated in isolation. This section deals mainly with the single technology element. The technology interaction effect is more complex in the case of three technologies and multiple stochastic variables. The discussion of this is deferred to Appendix 2, with the main conclusions reported here.

CO_2 price risk

In most of the cases shown in Figure 21, CO_2 price risk is not very significant. The exceptions are for gas and nuclear plants when a coal plant is on the margin. If a coal plant is at the margin of the merit order (left-hand block of results in Figure 21), CO_2 prices are assumed to be passed through to electricity prices at a rate determined by the high emission levels of a coal plant. Gas and nuclear investments would in this case be strongly affected by changes in CO_2 price. This is not the case for coal investments, since changes in CO_2 price would affect both costs and revenues by a similar amount, leaving overall profitability relatively insensitive to changes in CO_2 price.

When a gas plant is on the margin (central block of results in Figure 21), the rate of feed-through of CO_2 prices to electricity prices is significantly lower because of

FIGURE 21

Investment thresholds under different marginal plant assumptions

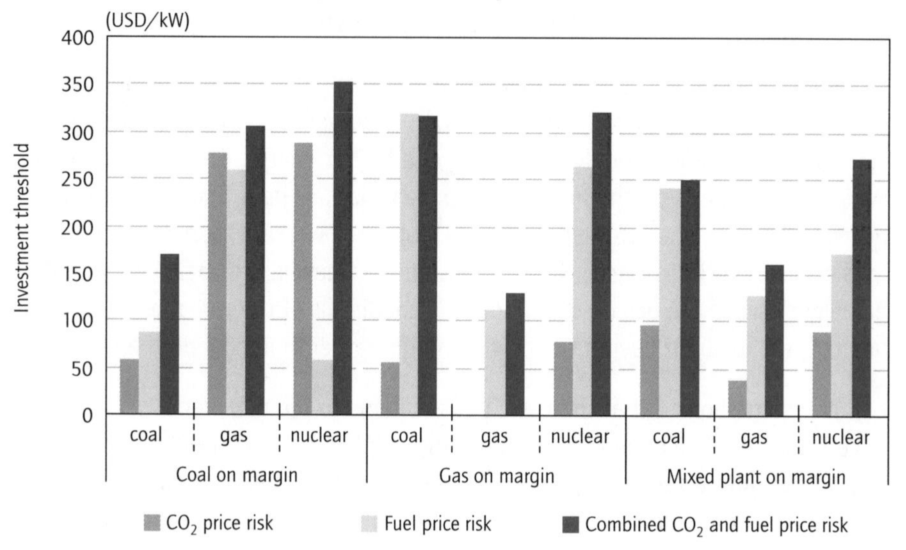

the lower emission levels of a CCGT plant compared to a coal plant. Therefore, the CO_2 price risk for coal and nuclear is quite low.

When the marginal plant is allowed to vary (*i.e.* a mixed plant on the margin as shown in the results on the right of Figure 21), the CO_2 price risk is still quite low as this case is closer to the 100% gas plant on the margin case than the 100% coal plant on margin case under the assumptions made in the model.

Fuel price risk

Coal prices are assumed in the model to be relatively stable, so fuel price risk is mostly created by uncertainty in gas price, and the possibility for this to feed through to the electricity price. In the case where a coal plant is always on the margin, the electricity price is unaffected by gas price uncertainty, so the fuel price risk for coal and nuclear plants is low. The fuel price risk for a gas plant, in this case, is high because gas price fluctuations would affect the generation costs without any corresponding change in the revenues.

In the case where a gas plant is always on the margin however, the fuel price risks for a new gas plant are low because fluctuations in fuel prices would show up in corresponding fluctuations in the revenue, leaving overall profitability relatively insensitive to fuel price changes. Coal and nuclear plants would be heavily exposed in this case to gas price fluctuations via the feed-through of these fluctuations to the electricity price.

The fuel price risk is reduced slightly in the case with a mixed plant on the margin because for some fraction of the time the coal plant would be setting the electricity price, thereby reducing on average the expected level of price fluctuations.

Combined CO_2 and fuel price uncertainty

The combined risk is not a simple addition of the fuel and CO_2 price risks, since it has to take account of the correlation between the two sets of prices. An interesting example is to compare gas and nuclear investments under the coal plant on the margin case. The nuclear investment has high CO_2 price risk but low fuel price risk, giving a combined risk threshold only slightly greater than the CO_2 price risk threshold. The gas investment has high CO_2 risks and high fuel price risks, but the combined total is not much higher than the sum of the individual components. This is because there is assumed to be some correlation between gas prices and CO_2 prices. Therefore, when gas prices are lower than expected (favouring the investment in gas), the CO_2 prices will also tend to be lower than expected, offsetting some of the benefits of the low gas price.

In any case, it is interesting to note that in all three cases of different assumptions about the marginal plant, nuclear investments appear to be amongst the most risky. This is fundamentally because CO_2 and fuel price uncertainties are expected to be reflected in electricity prices and will therefore directly affect nuclear plant revenues. For the two fossil-fuel technologies, fuel and CO_2 prices can affect both costs and revenues, which makes the profitability (difference between revenues and costs) less sensitive to fluctuations in these prices.

Technology interaction effect

The results in Appendix 2 indicate that technology interaction effects are of secondary importance compared to the investment thresholds presented in this Chapter. Whilst the different risk profiles for coal, gas and nuclear plants do provide some additional options value to having multiple choices available, these are only significant in the case of hedging the CO_2 price risk in the choice of coal versus gas (as discussed earlier), and tend to be insignificant when taken in the context of fuel price uncertainty.

Options for an existing oil plant

This example considers the early decommissioning of an ageing existing oil plant. We assume that the oil plant has a low utilisation rate, under average expected prices, that it is close to conditions where the plant would be standing idle and incurring fixed costs with minimal running time. It is assumed that the plant has 10 years worth of technical life, and would be due to be de-commissioned at that time at a total cost of USD 70 million. The early de-commissioning option would also cost USD 70 million, bringing forward this capital cost, but saving the fixed running costs of USD 6 million per year for the remainder of the plant's technical life.

Our final assumption is that the oil plant can respond to stochastic changes in prices by producing electricity when conditions are favourable, and going on stand-by with the normal fixed costs when conditions are unfavourable. This operational flexibility increases the value of the plant compared to simply taking average expected prices. The increases are shown in Table 3 for three different assumptions about price variation:

- CO_2 price variation only - geometric Brownian motion with 7.75% annual standard deviation with an additional jump in the range ± 100% in year 11.
- CO_2 price variation as above, plus fuel price and electricity variation modelled as geometric Brownian motion with 7.75% annual standard deviation. CO_2 and oil price fluctuations are not assumed to be correlated.
- CO_2 and fuel price variation as in the second point above, but with additional short-run volatility in CO_2, fuel and electricity prices modelled as an exponential mean-reversion process with annual volatility of 25%.

Early decommissioning is just about cost-effective if prices are deterministic, because although bringing the capital expenditure forward increases costs due to the time-value of money, this is outweighed by the savings made by avoiding the fixed costs. A DCF analysis would therefore imply that the decommissioning should be carried out immediately. In our results, we get a similar indication when only CO_2 prices are uncertain, as they do not create much of a difference to the cash flow. Under the first price example, we therefore get around 75% probability of early decommissioning (blue bars in Figure 22). By contrast, in the third of our examples, when we include fuel price uncertainty and short-run volatility, we get a more substantial stochastic value from the project, which outweighs the possible savings to be made from early closure - in this case the decommissioning is delayed until the final year of the plant. The second price example falls between the other two. Figure 22 shows this effect on timing of decommissioning.

TABLE 3

Value of operational flexibility under price uncertainty

	Additional present value of gross margin created by operational flexibility
CO_2 price uncertainty only	USD 3 /kW
CO_2 and fuel price uncertainty	USD 61 /kW
CO_2 and fuel price uncertainty plus short-run volatility	USD 135 /kW

FIGURE 22

Effect of uncertainty on the timing of plant decommissioning

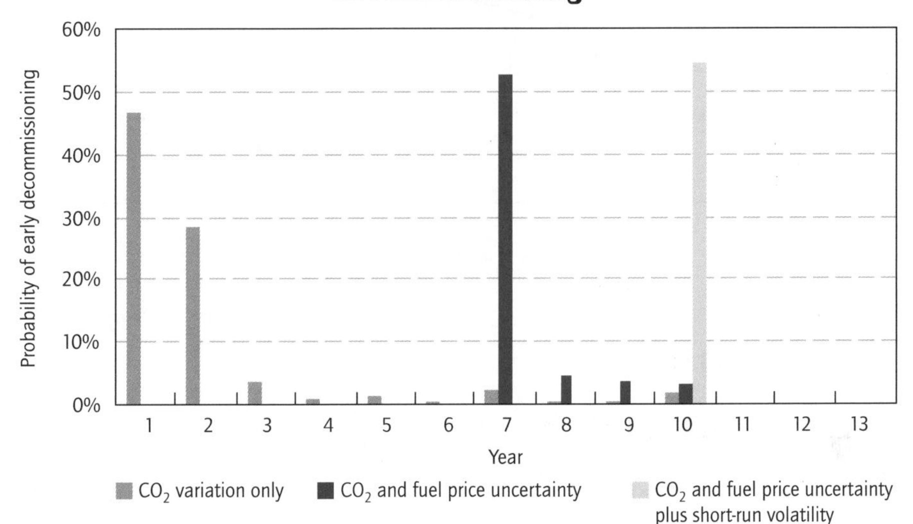

If uncertainty and volatility were taken into account, then decommissioning would tend to be delayed until the end of the plant's life.

The second investment option we consider for an existing oil plant is to replace the plant with a new coal-fired plant. It is interesting to compare this case with the earlier results where a new coal-fired plant was considered as a stand-alone investment. In that case, the option was to invest in a new coal plant or not to invest at all. In this example, coal is an alternative to the existing oil plant, which has quite a close emission factor under our assumptions of efficiency and fuel emission factors (0.70 tCO_2/MWh for the oil plant compared to 0.74 tCO_2/MWh for the coal plant). This means that the relative gross margin between a new coal and an existing oil plant is less sensitive to the price of carbon than in the stand-alone investment case for coal. This can be seen in Figure 23 where the investment threshold is slightly lower than was the case in Figure 11 for the stand-alone coal investment.

Figure 23

Investment threshold for replacement of existing oil plant with new coal plant

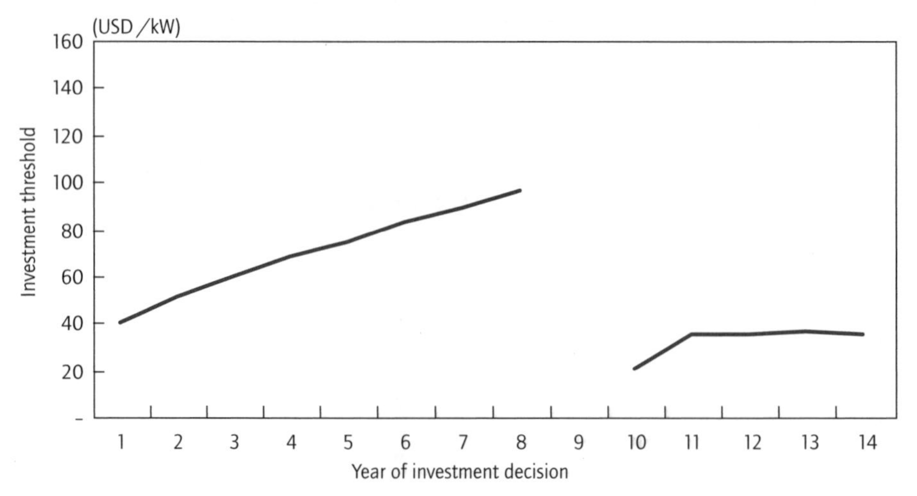

Policy options

Reducing price variability

There is a long-running debate amongst economists as to whether it is more efficient to achieve emission reduction by specifying the quantity of emissions to be reduced (such as in an emissions trading scheme) or by specifying the price of emissions reductions with an emissions tax (for a review of this literature see for example, Philibert, 2006; Pizer, 2002). A result of this analysis is that price instruments (taxes) are economically preferable when the marginal benefits of reducing emissions do not vary strongly with the actual level of emissions (as would be the case where there are no threshold concentrations that dramatically increase damages), whereas quantity instruments (trading) are economically preferable when damage threshold effects are important.

In this section, we look at whether the two instruments would differ in terms of the extent to which they create investment risk. We also look at the effect of other controls that could be placed on price variability.

So far, we have modelled carbon prices as having an annual variation as well as a "jump" in one particular year. The annual variation is meant to represent fluctuations in prices that might occur either in an emissions trading scheme or in other types of policy where the difficulty of meeting an emissions target might be uncertain (*e.g.* due to uncertainty in technical costs and projected emission levels). The "jump" in price is meant to represent a policy discontinuity, such as a change in policy regime or a significant change in stringency of targets. The uncertainty premiums shown so far therefore contain both the annual variability and the policy discontinuity elements, and could be considered to represent the case for an emissions trading scheme type policy.

For a tax, the carbon price would be known exactly as long as the policy was in force, so there would be no annual variation. This removes one element of the price uncertainty. However, companies would not be able to rule out the possibility of step changes in the tax level or changes in the policy regime, which might occur as a result of political review. The question then arises of how important the price "jumps" are compared to the annual variation – does removal of the annual variation significantly reduce the investment threshold? The answer appears to be no.

In order to quantify this effect, we simply switch off the annual price variation in carbon price, leaving the price "jump" element in place. The results are presented in Table 4 for two investment technologies where carbon price uncertainty has the strongest effect, namely carbon capture and storage and a nuclear plant investment when a coal plant is on the margin. The results are shown with only carbon prices being uncertain (*i.e.* fuel price uncertainty is switched off).

The effects of introducing a price cap (at 150% of expected price) and floor (at 50% of expected price) are also shown in Table 4. Note that the introduction of a symmetrical price cap and floor in this way would leave expected prices unchanged under our assumption that the price jump does not exceed +100%. In these runs, carbon prices have both an annual variation and a price jump component of uncertainty.

TABLE 4

Investment thresholds for nuclear and CCS under different price controls

Investment thresholds (% of capital cost)	Base case - jump and annual variation		Price cap (150%) and price floor (50%)	Jump only, no annual variation	Annual variation only, no price jump
	Case 1	Case 2	Case 3	Case 4	Case 5
	Price jump in year 6	Price jump in year 11	Price jump in year 6	Price jump in year 6	No price jump
Nuclear (with coal on margin)	31%	21%	28%	30%	13%
Carbon capture and storage	84%	37%	47%	80%	18%

The table shows the thresholds for a nuclear plant and carbon capture and storage investment options under different assumptions of price variability. The numbers in the table are calculated in terms of the additional net present value (expressed as a percentage of the project capital cost) that the project would need to exceed in order to stimulate immediate investment. These values can be compared to the hypothetical case where there is no price uncertainty, in which case there would be zero additional net present value required to stimulate investment.

The values of these investment thresholds are higher for CCS technology than for nuclear plants. This is because the CCS investment case is compared to a baseline of an existing coal plant. When the CO_2 price goes up, the CCS financial case improves whilst the alternative financial case (*i.e.* sticking with the unabated coal plant) gets worse. The relative difference between the CCS technology case and the unabated case is therefore more sensitive than in the case of a nuclear plant where the investment is considered against a fixed baseline of no investment. In absolute terms, the investment premiums expressed in USD /kW are more closely balanced.

From these results, it can clearly be seen that Case 5, the complete removal of the price jump, would have the greatest effect in reducing the investment threshold. However, this represents the case where some kind of guarantee could be given that no significant policy shifts would ever be made, which is rather unrealistic. Providing such an assurance for 10 years (Case 2) does mark a significant improvement compared to 5 years (Case 2), and is perhaps more politically feasible than Case 5.

Providing a price cap and floor (Case 3) does improve the investment threshold, but not as much as delaying the possible price jump (Case 2). Given our assumption of a flat probability distribution for the price jump, the cap would be expected to be exercised in 25% of cases and the price floor exercised in 25% of cases. Hence, there would be a 50% chance of some kind of price control under these assumptions. In the case of carbon capture and storage, the range between cap and floor would have to be narrowed such that one or the other would be in effect for 60% of the time in order to bring the investment threshold down to the same level as in Case 2, the policy extension case. Some of these results are shown in Figure 24.

Case 4 represents the pure tax situation (*i.e.* no annual volatility, but still the possibility of a jump in future prices). It can be seen that this does not significantly reduce the investment threshold compared to Case 1, indicating that the possibility of a price jump dominates the investment threshold, whereas annual variation is relatively unimportant. Therefore, based on the price volatility assumptions we have made, from the point of view of investment risk, taxes do not seem to perform any better than trading instruments, unless they can be linked to ways of providing more credible assurances against future price jumps.

We can look at the implications of a carbon price cap on investment incentives by examining how a price cap (without a price floor and keeping the targets unchanged) would affect expected future prices. We set up the carbon price process as before so that the expected carbon price is flat with 7.75% annual volatility (geometric Brownian motion) and a price jump of ± 100% in year 6. We then impose a price cap at 50% above the expected carbon price. This is rarely exceeded in the first five years of the run, but is quite often exceeded as a result of the carbon price jump. The effect of this price cap is therefore to limit the upper end of the range of carbon prices arising from the price jump in year 6 and beyond. The expected average gross margin of the low-carbon plant is reduced as a result of the price cap. The effect on a coal plant would be to improve the expected profitability (gross margin) of the coal plant by about 2.5% of capital cost, a rather weak effect.

FIGURE 24

Effects of carbon price controls on investment thresholds

Case 1: 5 years before price jump

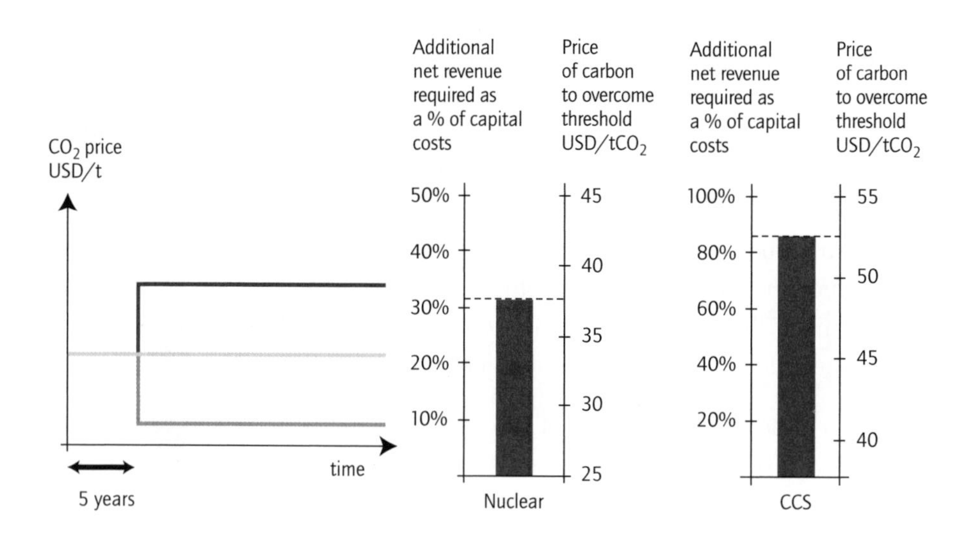

Case 2: 10 years before price jump

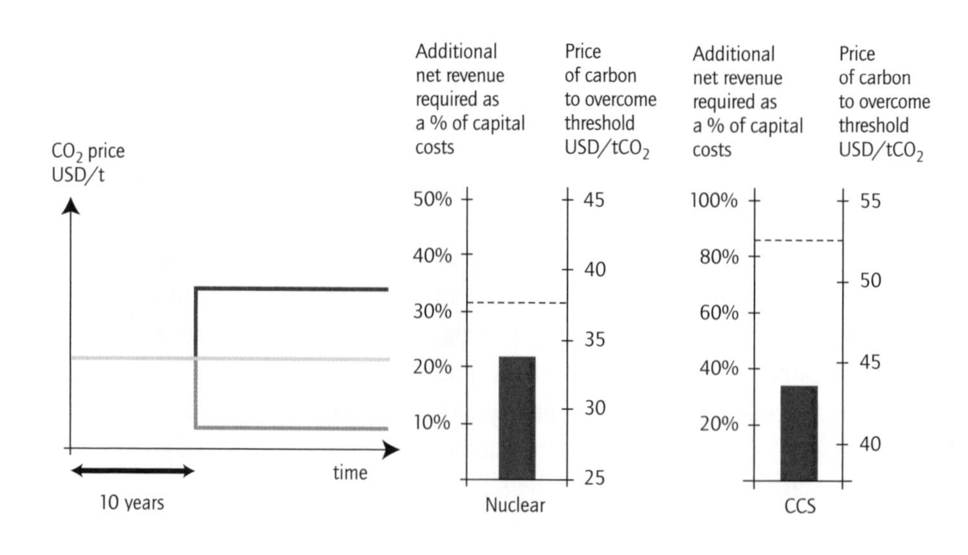

FIGURE 24 *(continued)*

Effects of carbon price controls on investment thresholds

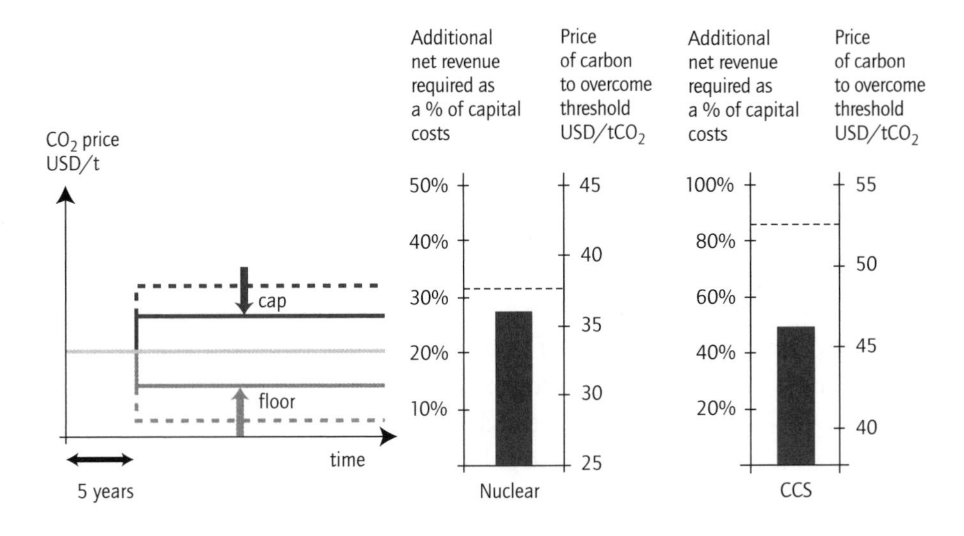

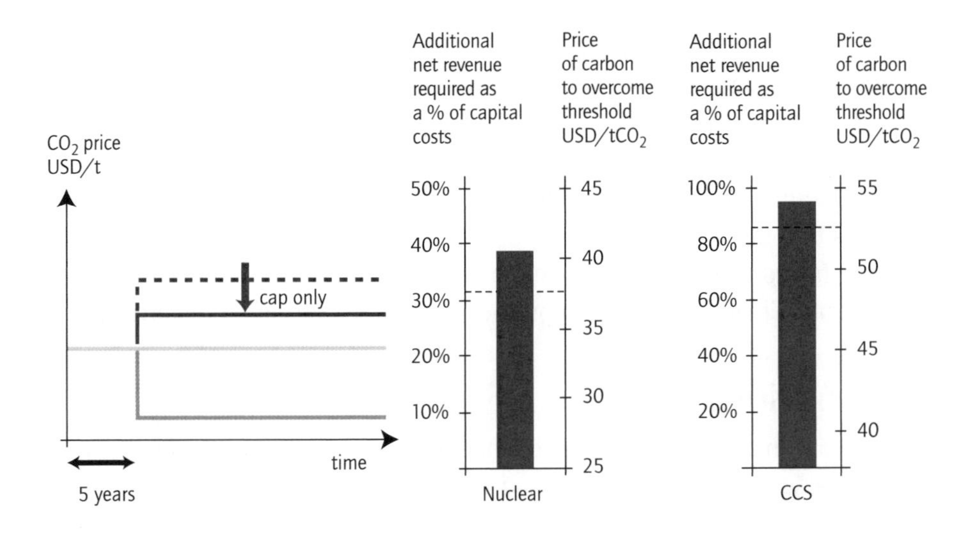

Stronger effects from price caps can be seen if we look at cases where the investments are more sensitive to changes in carbon price (Figure 24D). In the most sensitive nuclear *case* (where coal is assumed to always be on the margin), a price cap set at USD 37.5/tCO_2 (50% above expected prices of USD 25/tCO_2) would reduce the expected profitability of a nuclear plant by around 7% of capital costs assuming a price jump in year 6. In the case of retrofitting a CCS plant to coal, introducing a price cap of USD 57/tCO_2 (50% above expected prices of USD 38/tCO_2) would reduce the expected profitability of the investment by 12% of the incremental capital cost. These penalties would be weaker if the price jump occurred in year 11 instead of year 6 (5% and 8% of capital costs for a nuclear plant and CCS technology respectively).

It has been argued in Dixit and Pindyck (1994) that introducing price caps (without a price floor and keeping the targets unchanged) could have a perverse effect not only on the investment case, but also on prices themselves. Translating their arguments into the context of carbon prices, they argue that at the margin, introducing a price cap would incentivise building a higher emitting plant (or at least would be a disincentive for the building of low carbon technologies as seen above). By increasing the population of higher emitting plants, leading to greater emissions, this would lead to an increase in carbon prices and a greater probability that the price cap would need to be exercised. Indeed, if the effect were strong enough, it could create a dependence and vested interest in the continuing existence of the price cap in order to support plants that had been built with the price cap assumed to be in place.

The converse case (*i.e.* price floors without price caps) would have the opposite effect. Price floors would be expected to improve investment incentives for low emitting technologies by reducing the downside risk of carbon price collapse. The effect is symmetrical. In other words, a price floor set at 50% of the expected market price would improve a nuclear plant's profitability by 7% and a CCS retrofit profitability by 12% if a price jump were expected in year 6. This result assumes that there is no uncertainty about the level of the price floor beyond the six-year horizon. In practice, price floors themselves will also be subject to uncertainty. Creating certainty in practice is discussed further in Chapter 5.

Carbon price pass-through

An increase in carbon price raises the operating cost of many power generators in the market, and could be expected to result in an increase in the electricity price. As we have seen, carbon price pass-through acts as a significant buffer against carbon price risk, although will not exactly cancel out the risk.

In our earlier results for coal plants, we assumed that 100% of the carbon price is passed through to electricity at a rate determined by the marginal plant. Any plant with a higher emission factor than the marginal plant is disadvantaged by an increase in carbon price, whereas any plant with a lower emission factor is advantaged by an increase in carbon price.

We can test the importance of our assumption about carbon price pass-through by varying this percentage pass-through rate. In this example, we set the rate of carbon price pass-through to 50%. This means that when carbon prices jump in year 6, there is a much lower corresponding jump in electricity prices. The effect of this on a CCGT plant is to reverse the relationship between the gross margin and carbon price - an increase in carbon price now reduces the gross margin slightly.

For a coal plant at 50% pass-through, the sensitivity of the gross margin to a change in carbon price is twice as high as in our base case. A 10% increase in carbon price leads to a reduction in the gross margin by 7% (compared to a reduction of 3.6% in our earlier results for a coal plant). A coal-fired plant would be more strongly exposed to carbon price uncertainty in this case as can be seen in Figure 25.

Allowance allocation

Closely linked to carbon price pass-through is the issue of free allocation of allowances in an emissions trading scheme. The EU emissions trading scheme directive allows provision of a "new entrants reserve" whereby free allowances can be allocated to companies wishing to build new plant. The reason for making this provision was to prevent possible distortions arising between incumbents who generally have a high proportion of allowances allocated for free and new market entrants. There is also a competitiveness issue relating to the fact that if one country in the scheme provides free allowances to new entrants, then other countries will want to follow suit in order to avoid the perception that they are a less attractive destination for investment.

From a standard economics perspective, free allocation of allowances is equivalent to a transfer of assets to the company - it affects the asset value of the company, but not the operating costs. This is because once the company owns

FIGURE 25

Effect of carbon price pass-through rate on investment threshold for coal

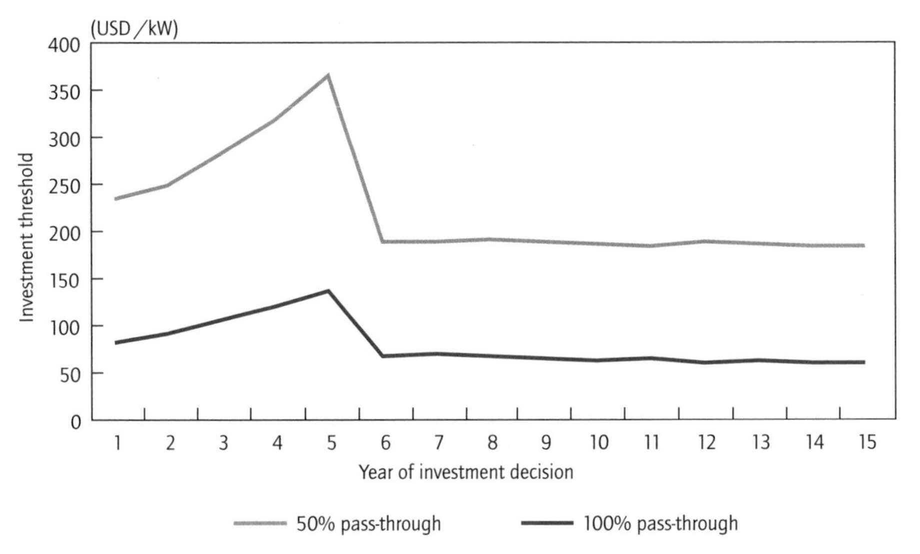

Reducing the pass-through of carbon price to electricity price from 100% to 50% significantly increases the investment threshold for a coal plant (assuming the full costs of carbon are borne by the generator in both cases).

them, allowances have an equal value (determined by the carbon price) whether they were allocated for free or not, and therefore the opportunity cost of using them to cover emissions is the same as the real cost that would be incurred if the company had to buy them.

Viewed from this perspective, a free allocation on the basis of historical emissions acts as a subsidy, the size of which depends on the emission levels of the technology being considered. For example, 10 years worth of allowances covering 90% of emissions at USD 20/tCO_2 would be worth approximately USD 1.0 billion for a new coal plant, and approximately USD 0.5 billion for a new CCGT plant. Leaving all other parameters unchanged, introducing free allocation on this basis would therefore substantially benefit a coal compared to a gas plant.

For a project to experience an increase in value arising from free allocation, it would still be necessary for the carbon price to be passed through to the electricity price. As an alternative, we can consider a variation from the normal economic perspective – where only the direct cost of additional purchased allowances needed for compliance with the emission target is considered to be a cost, and opportunity costs associated with freely allocated allowances are not taken into account. This could be the situation for example for a price-regulated utility.

In this case, if we assume that 90% of the allowances are allocated for free, then the direct cost of carbon is only 10% of the CO_2 costs in our base case. For a coal generator, if the CO_2 price were USD 20/tCO_2, electricity prices would increase by around USD 2/MWh or 7% if only 10% of the carbon cost was passed through to the electricity price. This would increase to USD 4/MWh or 14% for a 20% pass-through based on an 80% free allocation.

This type of compliance-driven behaviour would also mean that variation (uncertainty) in carbon price would have a much-reduced effect on the gross margin, since carbon price represents a smaller fraction of the project costs and revenues. A 10% increase in carbon price now only reduces the gross margin for the coal plant by 0.7% and leaves the gross margin for the gas plant virtually unchanged. We therefore expect that with this type of compliance-driven behaviour with low direct costs and low carbon price pass-through, uncertainty would have a much less important effect on project profitability, and the investment premium arising from uncertainty would be much less significant than seen in the base case. In fact, when variations are so low, it is difficult to detect the effect on investment threshold with the model, so no results are shown here.

Volatility versus uncertainty

There are two aspects to price uncertainty, a short-run volatility element, which is expected to oscillate around some mean value, and then a longer-run uncertainty about the value of this mean. We have made the case that we should only consider the longer-run uncertainty, since short-term oscillations around a mean will not affect average investment conditions. In other words, we have assumed that short-run volatility does not increase the value of waiting because it does not yield valuable information about expected future investment conditions.

However, this assumption omits one potentially important effect. Power plants may be able to benefit from short-run volatility by optimising their dispatch decisions - choosing not to run the plant when price conditions are unfavourable. Considerations of plant operating economics suggest that this operational flexibility is likely to be more significant for a plant with low capital and high operating costs (*e.g.* gas) than for a plant with high capital and low operating costs (*e.g.* coal). We can test this by introducing a short-term mean reverting volatility of 25% on top of the longer-term uncertainty in price, and look at the effect on the range of the gross margin experienced by the plant taking into account this operational flexibility. The results of these tests can be seen in Figure 26 for coal and Figure 27 for gas.

FIGURE 26

Coal plant cash flow under different fuel price volatility assumptions

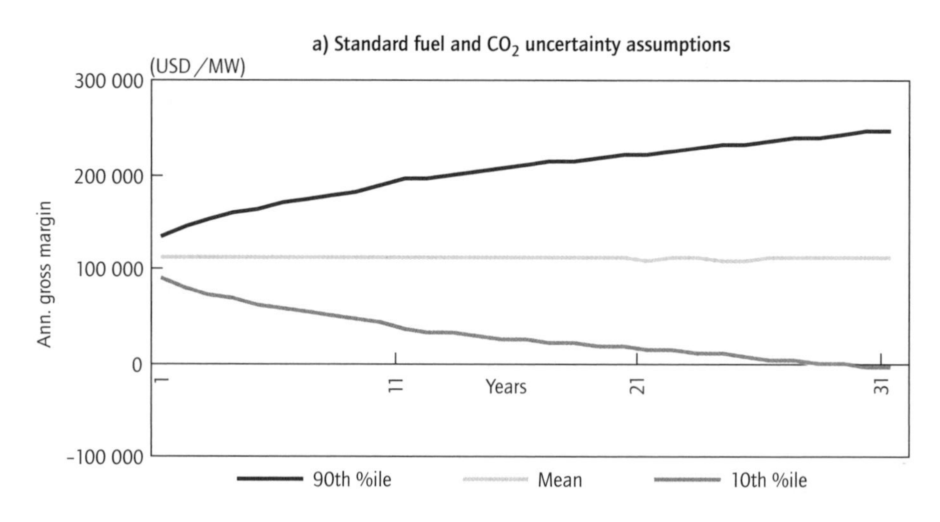

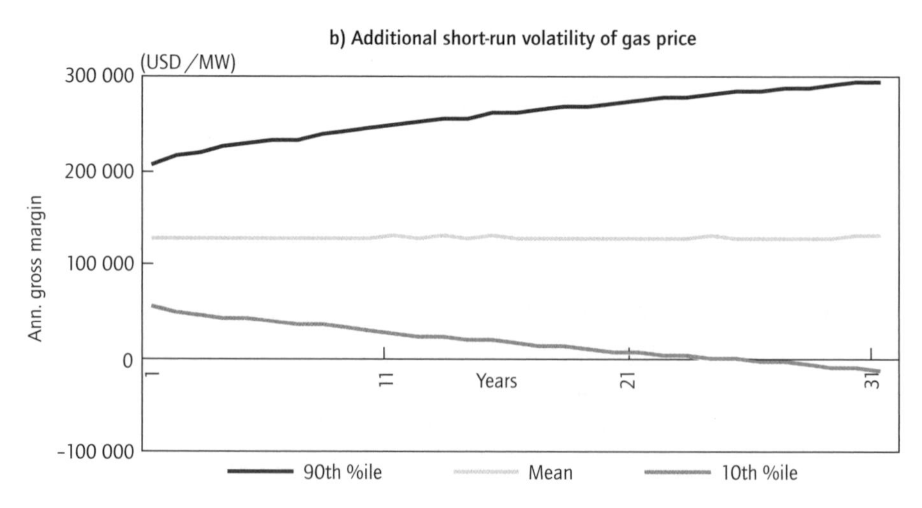

FIGURE 27

Gas plant cash flow under different fuel price volatility assumptions

a) Standard fuel and CO_2 uncertainty assumptions

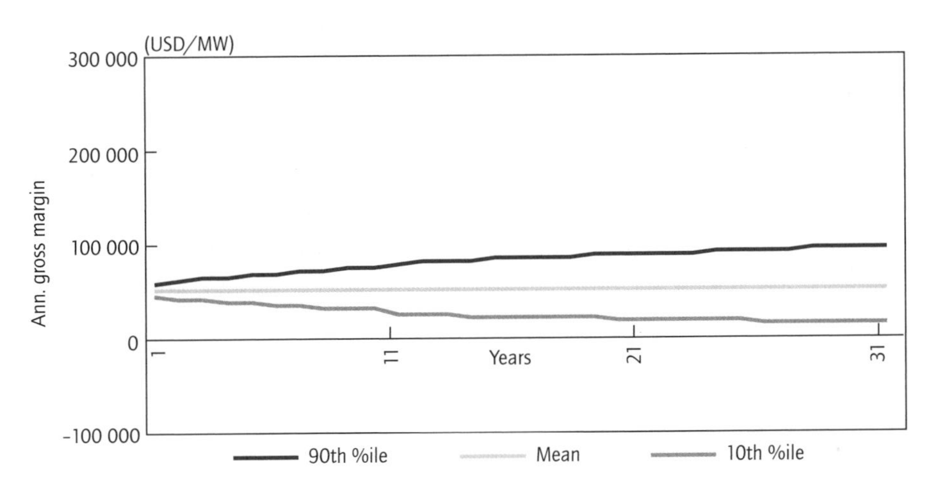

b) Additional short-run volatility of gas price

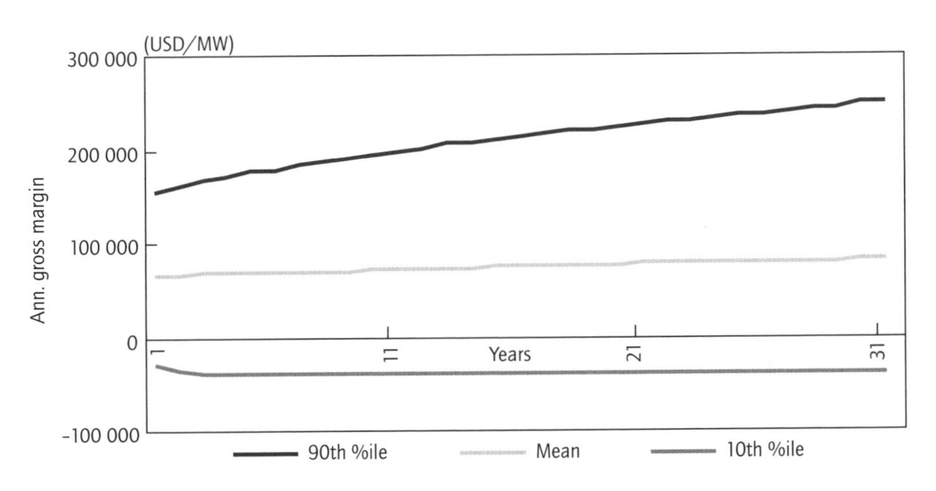

As for our earlier examples, the fuel and CO_2 price assumptions are chosen to make the NPV for the gas and coal plants equal. Since the capital costs of a coal plant are higher than for gas, coal operates with a higher gross margin. This means that when the high short-run price volatility is included, the gross margin for a coal plant is less likely to go negative than for a gas plant (compare Figure 26b and Figure 27b). For the gas plant, the ability to switch off during periods when the gross margin goes negative effectively caps the losses at the level of the fixed operating costs for the plant. This leads to a slight rise in the mean annual gross margin in Figure 27b, raising the overall project NPV by around USD 40/kW, approximately 9% of capital cost. In the high volatility case shown here, we have assumed a lower value of correlation between gas and electricity price (20% correlation factor), which tends to increase the effect of volatile prices on the gross margin because revenue and cost variations are no longer strongly coupled. This assumption of lower correlation is quite plausible for short-run volatility even if long-run prices of electricity and gas were expected to be closely matched. If we extended our standard assumption of a high correlation between gas and electricity prices to the short-run volatility, then price variation would have a much lower impact on the gross margin, and the effect of operational flexibility would be negligible.

There are two other competing factors that are not captured by these results, which could alter the size of the effect of volatility on project value. The full range of risks that a plant might be subject to during its life is not captured here. There may be other risk factors that are stronger than fuel price risk, so that operational flexibility could become more important than indicated by these results. In addition, we have assumed that the upside benefits of price volatility will remain un-capped. This may not be the case if extended periods of a high gross margin are seen as excess profit and are regulated away (*e.g.* through additional taxation). If this were seen as a potential risk by the investors, then volatility would not add to project value in the way shown in these results.

COMPANY PERSPECTIVES ON MANAGING RISKS

Clearly, the real world is a great deal more complex than the model used to produce the results presented in Chapter 3. An important part of this project has been the interactions with companies involved in power sector investments to gather views on how risks are dealt with in practice. These are incorporated into the discussion in this Chapter. Interactions with companies in this study include a combination of dedicated workshops held at Chatham House, company meetings and telephone interviews. A list of companies consulted during the course of this study is included in Appendix 3.

A key feature of the modelling approach used in this study is that it considers all prices to be exogenous to the investment case, so that investors are assumed to be pure price-takers. Individual projects are treated in isolation from the market environment in which the project is being considered (for example, the analysis ignores any feedback that a large power generation project might have on electricity prices, and ignores any impact that one company's decision might have on its competitor's investment decisions).

In reality, market dynamics are important; investment decisions and strategic behaviour can be strongly influenced by the type of market in which the company is operating. A monopoly provider or a company with regulated prices will have different investment drivers from a company operating in a competitive pricing environment. This market context is discussed in the first section of this chapter.

Likewise, we made various simplifying assumptions in our representation of climate policy uncertainty. The second section examines companies' expectations regarding climate change policy uncertainty, and explores differences with the assumptions made in the model. The third section discusses some of the other risk factors that companies face in the investment decisions that have not been explicitly modelled. The chapter concludes with a discussion of how companies respond to risks.

Importance of market structure

Investment drivers in different market types

Decisions to build new plants may be made in response to various drivers, including: an expected increase in demand; replacement of an existing plant due for retirement; and/or for strategic reasons, to enter new markets. These drivers may depend on the structure of the electricity market. The structure of electricity markets varies considerably across IEA countries, with a range from liberalised-competitive markets through to price- regulated markets and markets with regionally dominant companies. Many countries in North America, Europe, Japan and Australia operate with a mixture of different electricity market types, sometimes as a result of an ongoing transition from regulated to liberalised markets in that country, other times as a result of different sub-national markets being regulated in different ways.

The investment drivers for companies will be different in different types of market. The context into which climate change policy is being introduced is not necessarily a smooth predictable one. The complexities exist independently of the introduction of climate change policies, and may continue to dominate investment incentives irrespective of climate change policy uncertainty.

In competitive markets, producers receive a signal to invest through the product price. When supply is becoming tight relative to demand, prices should rise creating the incentive to invest in new capacity. Because of the time taken to bring a new power plant online, this process requires some judgement in advance of likely impending shortfalls in the market. Timing of investment can be critical. White (2005) describes how price behaviour in competitive markets will tend to spend periods at low prices (close to short-run marginal cost), which are too low to encourage a new entry. As a plant retires or demand increases, the market gradually becomes tighter until prices spike up above the threshold for a new entry. At this point, there is a race to bring a new plant on-line to make the most of the higher prices. The period of higher prices leads to additional capacity in the market, which once again returns the market to a period of low prices and low investment until the next price spike. The price behaviour is shown schematically in Figure 28, following the form in White (2005).

Figure 28

Investment and price cycles in a competitive market

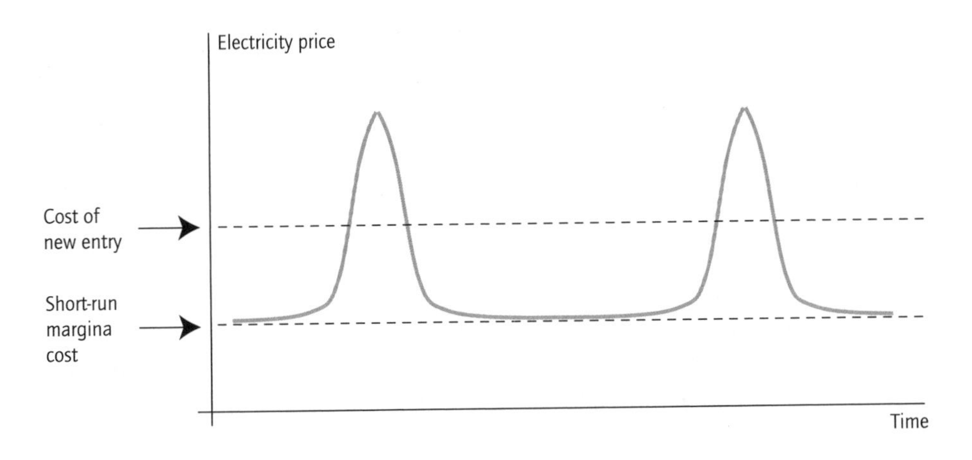

Such pricing behaviour for a competitive market can be considered as a dynamic equilibrium as long as a sufficiently long time perspective is taken. A mathematical treatment of this behaviour is given in Dixit and Pindyck (1994). However, this type of herding behaviour creates challenges both for companies and for policy makers. Companies need to judge their investments carefully with respect to the occurrence and duration of price increases in order to ensure they benefit from the periods of higher price. It also challenges policy makers, who have to hold their nerve that the market will deliver the required investment to meet demand despite price spikes. An expectation of policy intervention in the market will create additional uncertainty and in the long run create additional barriers to investment.

In most real markets, these boom and bust cycles are dampened in a number of ways, each of which effectively gives companies additional market power compared to the pure competitive market situation. There is a degree of vertical integration, such that the majority of power generation companies also own retail/supply businesses. Whilst most supply businesses are not protected from competition, this arrangement does provide something of a buffer, at least against shorter-term market variations, and increases a company's ability to plan capacity additions.

In addition, many electricity markets operate as oligopolies or monopolies with a small number of - possibly only one - large incumbent player(s) having considerable market power. Companies in this position have much greater control over the market dynamics, and will be able to plan capacity additions in a smoother manner consistent with expected changes in demand - supply balance. Electricity prices may still be determined through a competitive bidding system (*e.g.* through a pool or other similar mechanism), but price will not be the only driver of investment. Investment incentives in these situations will also be driven by considerations of market share and cost minimisation of the portfolio.

Alternatively, prices may be set by a regulator. In these circumstances, prices will be determined according to a formula that will usually allow recovery of agreed operating costs and an agreed rate of return on new investments. As for any company, the key priority for the power generators will be to maximise return on investment, but this return will be largely determined through a negotiation process between the company and the regulator. Maintaining good relationships with the regulator, developing joint visions of the required rate of capacity expansions and technology requirements, as well as cost minimisation therefore become key investment drivers for companies in this situation.

This last case of electricity price regulation contrasts strongly with the assumptions we have made in the modelling work. In the model, we assumed that investment decisions respond to changes in electricity price signals, whereas in price regulated markets, it is the other way round - electricity prices will respond to investment decisions. This fundamentally alters the risk profile for investment decisions.

Impacts of risk on electricity price and investment levels

Most of the companies interviewed during the course of this study generally agreed with the overall conclusion that power generation investments will ultimately be driven by their view of electricity and fuel prices. As far as the basic choice between coal and gas plants for base-load generation goes, electricity and fuel price risks in most markets are probably stronger than climate policy risks. Nevertheless, the general perception is that climate policy risk is increasing in importance.

The most likely response to an increasing level of risk will be to favour investment in low capital cost options. These will represent the "least expensive mistake" if future conditions turn out to be less favourable than expected. The lowest capital cost option may be to extend the use of an existing plant beyond their normal technical lifetime, allowing companies to delay the replacement of an existing plant until more information is obtained. If capacity expansion is required to meet expanding demand, the lowest capital cost option for new generation is a gas-fired plant, and this may gain a competitive advantage over other technologies in a more uncertain investment environment.

Increasing the level of risk in the electricity sector will create a requirement for higher rates of return, which means that electricity prices would need to rise. There are two mechanisms by which such price rises might occur. First, the cost of entry to the market should rise to incorporate the increased price of risk (both in the debt and equity components). If the cost of entry increases, the electricity price would rise accordingly, since in a competitive market the long-run electricity price will be determined by the cost of entry. In this situation, the same level of investment would be achieved, but at a higher equilibrium price of electricity. The second mechanism by which electricity prices might rise is if the reserve margin drops (*i.e.* reducing the levels of spare generation capacity in the system).

Figure 8 gave an illustration of how an electricity price is derived in a competitive market based on balancing demand at any given time with a supply stack, which is dispatched in order of increasing marginal cost. If electricity demand grows over time (*i.e.* the demand curve in Figure 7 is shifted to the right), and new capacity is not added to the system, then a generation plant further up the supply stack will get dispatched more often, leading to higher electricity prices. As indicated in Figure 7, the supply cost curve is often steep towards the right hand side, so the required electricity price increases could occur with relatively small decreases in the supply margin. In that example, the price would increase at a rate of over USD 3/MWh per percentage point reduction in reserve margin. The level of risk premiums identified in Figure 10 may therefore be covered by a reduction in the reserve margin of less than 1%.

Which of these two mechanisms occurs in practice probably depends on the market design. In some markets, there are explicit mechanisms for maintaining reserve margins in order to ensure certain standards are achieved for the reliability of supply. This can include explicit mechanisms to support investment in reserve capacity or by simply applying reliability standards. In these cases, reserve margins may not drop, but investment may be steered more towards peaking and investing in a mid-merit order plant than would be considered optimal in a more certain investment environment. Again, this would tend to favour investment in a gas generation plant over base-load coal or nuclear generation investments.

How is the value of waiting affected by market structure?

The investment thresholds presented in Chapter 3 are based on the value of waiting that arises when there is a source of uncertainty that could be resolved at some time in the future. We looked at two sources of uncertainty:

- Fuel price uncertainty that is gradually resolved over time (since each annual change in price is assumed to bring a new expected value of future prices).
- CO_2 price uncertainty that combines a gradual resolution of price uncertainty each year, together with a more abrupt resolution of uncertainty relating to a policy change at some fixed point in time in the future of the project.

However, the question arises of whether the option value of waiting is maintained given the competitive and/or strategic investment drivers that companies face.

In theory, a monopoly generator would be in a position to choose the optimal timing of their investment, taking into account the possibilities to resolve any future uncertainties by waiting. The option values calculated in Chapter 3 should, in principle, apply directly to decision-making considerations of monopoly companies (Dixit and Pindyck, 1994). In this case, companies often have considerable flexibility to defer investment, perhaps temporarily extending or expanding output from an existing plant during the deferment period.

At the other extreme, it can be shown that in a competitive market, the threat of entry into the market by other players raises the cost of waiting. This is because entry by a competitor may adversely affect the business case for new investment. In the limit where there are many possible new entrants to the market, the real options value could be competed away (Trigeorgis, 1991). This is why the signalling of planned investments by incumbent players is often used to try to reduce the attraction of a new entry.

In an intermediate case where there are a few players in the market, it has been shown that the real options value is reduced, but is not entirely competed away (Lambrecht, 2003). The investment threshold would be reduced by a factor representing the probability that a given company "wins the race" to be the first to invest. In the case where there are an infinite number of companies investing, the probability of being the first goes to zero, and the option value is entirely competed away.

The initial answer, therefore, seems to be that the option values presented in this study would only apply in markets where there is appreciable concentration of market power, and that in fully competitive markets the normal NPV rule would be re-established since management effectively does not have the flexibility to respond optimally to the uncertainty in question. However, there are some caveats to this answer.

To begin with, the first-mover advantage that creates the incentive to be the first to invest in a competitive market is a response to an opportunity to earn revenues, mainly driven by electricity prices as described in the previous section. However, the climate change risk will not be affected by the order in which investment occurs (*i.e.* the first-mover's investment will still be subject to the risks created by the policy uncertainty). Therefore, the second-mover advantages of waiting and capitalising on the possible mistakes of the first-mover may be just as strong as the first-mover advantage. In any case, first-mover advantages in the electricity sector are probably not as strong as in markets such as electronic goods and retailing where companies may be able to gain a dominant position through exploiting brand loyalty and imposing their own technology standards. It is therefore likely that the risks resulting from the policy uncertainty would persist even in a fully competitive situation.

How do climate change risks affect the cost of capital for a company operating in a competitive market? As was discussed in Chapter 3, the CAPM model suggests that an increase in systematic risks would lead to a higher cost of capital, whereas an increase in diversifiable risks should leave the cost of capital unaffected. On the one hand, climate change policy uncertainty might be considered diversifiable, since there will be winners as well as losers arising from unexpected changes in climate policy. On the other hand, given the basic nature of energy consumption to the economy, climate change policy may be fundamental enough to affect the market as a whole, and may therefore be considered to be a systematic risk. In this case, the additional risk created by climate policy uncertainty would feed into the investment decision via an increased cost of capital and have a similar effect to the increased thresholds presented in Chapter 3.

In practice, it is difficult to determine precisely how these factors will play out in different markets, and whether climate change policy risks would be treated by financial markets as systematic or diversifiable. In actively managed financial portfolios, investors may view climate change as a non-diversifiable risk because the inability to predict how climate change policy might affect any individual company's performance. Thus, it is not possible to create a portfolio of assets which fully diversifies away the risk. So far, the risks have not been considered sufficiently and have not provided an incentive to acquire greater understanding of how the risks will be overcome compared to other risks faced by power companies.

In any case, the CAPM model itself has some limitations in explaining actual investment behaviour, since it represents equilibrium for the market as a whole. Deviations from this equilibrium in the case of individual firms or projects may be quite significant. For example, risk aversion on the part of power companies may still play a considerable role in decision making, leading to a greater level of hedging than the traditional CAPM model would suggest. Several power companies interviewed expressed an underlying tendency to want to diversify their portfolios by generating assets across different generation technologies to "spread the risks", even though diversification by individual companies is not required to minimise risk in the financial markets as a whole. Individual decisions will incorporate considerations of the risk of bankruptcy and the need to avoid the possibility and costs of financial distress.

In summary, the presence of climate change policy uncertainty is likely to be factored into companies' investment decisions even in competitive markets, leading to the addition of risk premiums along the lines we have presented in this report.

Climate change policy risks in the real world

Climate policy uncertainties faced by companies are considerably more varied than the way they are represented in the model. Greater complexities include the relationship between carbon price and other commodity prices, the way in which uncertainty will be resolved, and details of policy design not relating directly to prices.

The model provides a representation of this relationship based on an emissions-trading type policy, with prices determined in a competitive market setting. Specifically, CO_2 price fluctuations are closely correlated with gas price fluctuations, and electricity prices are determined by the short-run marginal costs of the marginal plant in the merit order. In practice, these relationships will depend on the type of market in which a company operates.

Many power companies operate under price regulation, where the price they receive for their electricity is set according to a formula applied by a regulator. These formulae vary, but typically, they allow generators to recover the ongoing costs of generation (*e.g.* fuel, operating and maintenance costs), and a return on capital investment costs. The cost recovery would likely include any costs of meeting environmental regulation. The cost of carbon would still be incorporated into the price of electricity, but this would be done on the basis of the average cost increase across the company's generation portfolio. For a company with a mixed portfolio (*e.g.* nuclear, hydro, gas, coal and renewables), only the cost increase for the fossil-fuel fired portion of the generation would be passed through to the electricity price. Since the operating costs for the zero-emitting plant would not be affected by introduction of a carbon price, the cost-recovery for this portion of the company's portfolio would be unchanged. This means that the influence of carbon price uncertainty would be weaker in a price-regulated market than shown in our results, which are based on a competitive market pricing model.

That is not to say that CO_2 policy risk for price-regulated companies is necessarily insignificant. Price-regulated companies negotiate the price formation process with their regulators, and recovery of the full additional costs of climate policy compliance is not guaranteed. In general, environmental compliance is treated as a necessary cost, but if costs escalate due to unforeseen circumstances (*e.g.* an increase in the price of international emissions credits), then there may be limits to the extent to which rate rises can be negotiated to cover these. Policy uncertainty makes such investment decisions more difficult.

In the model, we assumed that there is a close correlation between the annual fluctuations in the price of gas and the annual fluctuations in the price of CO_2 (not including the jumps in CO_2 price representing policy uncertainty). This assumption is made in order to represent the fact that the switch between coal and gas will be a key abatement option in the power sector. In effect, we are saying that there is a local equilibrium in the price of CO_2 caused by the expectation that there will be a large supply of allowances at this price due to the availability of this fuel switching option. The actual price at which the switching occurs depends on the gas price, but it is a technologically stable solution over a wide range of abatement targets. There may be other such price "equilibria" caused by the presence of other technology options (*e.g.* renewables, nuclear, carbon capture and storage) as indicated schematically in Figure 29.

FIGURE 29

Schematic CO_2 abatement cost curve

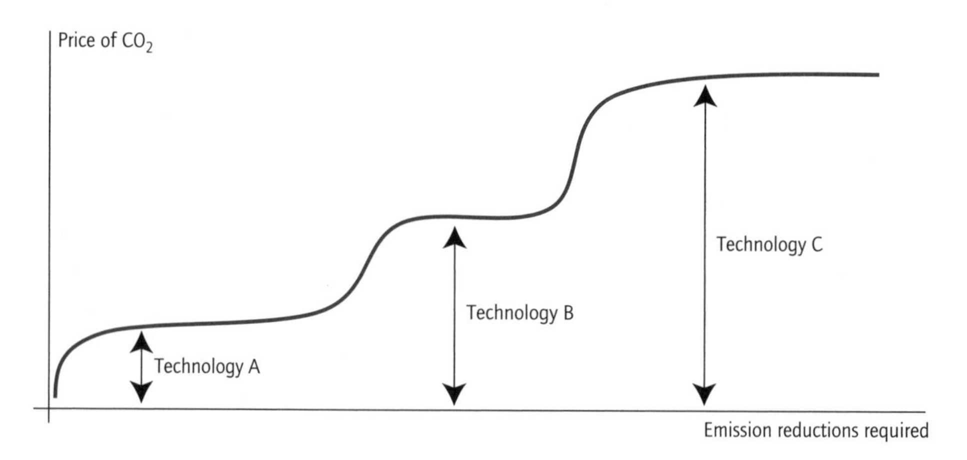

In a closed emissions trading scheme, the price of carbon may tend to take on discrete values determined by the costs of different abatement options.

In this respect, our assumption that policy uncertainty leads to an equal probability of any future price within a certain range may not accurately reflect real outcomes. This may be an important effect in a closed scheme (*i.e.* where offset projects and international credits are not allowed in significant volumes), and where the price of carbon may be dominated by the relatively few and homogeneous abatement options available to the power sector.

If the abatement cost curve is "lumpy" in this way, then long-term carbon prices may be relatively stable to small changes/uncertainties in abatement levels of an emissions trading scheme in some regions, and highly sensitive to small changes in other regions (*i.e.* close to where the major introduction of a new technology would be needed to meet the target). This price response of CO_2 to the ambition level of the policy is rather more complicated than we have modelled and may warrant further investigation.

In the model, we assumed that annual variations in gas prices and CO_2 prices are correlated, but that the possibility of a policy-driven jump in CO_2 price does not affect the gas price. It is possible, however, that in real energy markets, a significant shift in the stringency of CO_2 targets could feed through to the price of gas. This could occur if there was a significant change in demand for gas as a result of the policy change, for example as a result of switching away from or towards coal. For this effect to be important, three conditions would need to apply:

- Climate policies that are not technology-specific, so that compliance could be achieved in large part through switching fuel.
- A sufficiently large potential for switching to gas from other types of generation (or *vice versa*) as a result of a shift in climate policy.
- A sufficiently high price elasticity of demand in the gas market so that the change in gas demand is reflected in medium to long-run gas prices.

If sudden changes in climate policy were to feed through to the gas price, this would increase the "gearing" effect of climate policy uncertainty on investment decisions, increasing the effective risk premium. The model results for the effects of fuel price uncertainty give some indication of the additional scale of the effects. For a more detailed assessment of the size of this effect, a more in-depth analysis would need to be undertaken on the effects of climate policy on gas demand and the effects of changes in gas demand on medium to long-run gas prices.

The results in Chapter 3 showed that the effects of climate policy uncertainty depend strongly on which type of plant is at the margin of the merit order, setting electricity prices. If coal is on the margin, CO_2 price risks are significantly greater than if gas is on the margin because of the greater rate of CO_2 price feed-through to electricity price (assuming competitive market pricing for electricity). In principle, if fuel switching from coal to gas is expected to be a significant abatement option in an emissions trading scheme, then prices would be expected to settle at a level just high enough to encourage operation of gas plants in preference to coal plants (*i.e.* bringing gas plants earlier in the merit order than coal plants). This would mean that closed emissions trading schemes with sufficiently stringent targets would tend to result in electricity markets operating with gas being dispatched before a coal plant in the merit order. Less efficient open-cycle gas turbines could still be used for peaking in this situation.

In the model, there was not a hard link between CO_2 price and the merit order, allowing coal, gas or a mixture of coal and gas to operate as the marginal plant. To some extent, this reflects what has been observed in practice in the EU emissions trading scheme; although the CO_2 price has been closely correlated with the gas price, it has rarely been high enough to actually trigger a switch from a gas plant on the margin to a coal plant on the margin on a regular basis. Broader geographical coverage of emissions trading schemes and inclusion of project credits such as Joint Implementation (JI) and Clean Development Mechanism (CDM) mean that there is less likely to be such a hard link between CO_2 price and merit order.

Another simplifying assumption we made in the model is that the information associated with the price "shock" is revealed all at once. This probably exaggerates the value of waiting, since in reality, information about future policy and price changes would be revealed gradually, with market players gaining some information before the uncertainty "event", and probably still gaining information and establishing suitable market prices for some time after the event.

In addition, the type of information revealed will be much more heterogeneous. We have used CO_2 price as a proxy for climate change policy, but the actual impacts of policy are more varied. In an emissions trading scheme, participants will be concerned by the many details of the scheme design, including: the overall abatement target; free allocation levels and methodologies; the allowance rules and availability of project mechanism credits, including JI and CDM credits; rules for new entrants and plant closure; and the possibilities for linking to other emissions trading schemes, as well as many other details that can have important effects on project and company profitability.

In competitive markets, free allocation to power generators tends to increase their asset value, since the full opportunity cost of CO_2 emissions for the marginal plant in the system would generally be passed through to the electricity price. For price-regulated companies, however, the costs of generation (including the costs of compliance with climate policy) are more transparent to external scrutiny, since they are an important part of the rate negotiation with the regulator. Price-regulated companies would therefore expect to only pass through the actual cost – the number of allowances purchased from the market – not the opportunity cost. This potentially makes price-regulated companies more sensitive to the levels of free allocation, as there may be some risk that not all climate policy compliance costs will be allowed to be recovered through the rate if free allocation levels turn out to be low and prices turn out to be high.

With a tax mechanism, the policy would effectively determine the carbon price level directly. This price could be completely independent of other commodity prices and technology costs, making it an apparently simpler policy mechanism. In practice though, there could be some form of recycling of tax revenues back to the affected sectors, so that the actual financial impacts of the overall policy would be more complicated than the simple price determined by the level of the tax. In any case, as has been argued in this report, a tax is not immune to a change in the level by subsequent governments, and so would still be subject to the sort of price shock event that we have modelled.

Other policy mechanisms will have their own particular patterns of uncertainty. For example, technology standards would essentially impose the technical uncertainty of the required project on the generators, which may come in the form of uncertain capital costs, uncertain plant performance or uncertain operating and maintenance costs. Further discussion of technical risk is given below. Capital grants for the early stages of new technology deployment may be a relatively certain source of income for individual projects, although uncertainties over the duration of capital grant schemes may affect firms' views of market potential and therefore willingness to invest in product development.

In summary, it can be concluded that in real emissions trading schemes, there may be several factors that limit the variability of CO_2 prices compared to the assumptions used in the model, and that the effects of uncertainty indicated by the model may therefore be an upper estimate. Other types of climate policy will not have the same degree of interaction with other commodities as assumed in the model. They will nevertheless be subject to uncertainty of different sorts that will act to raise investment thresholds.

Uncertain timing of policy introduction

In the analysis presented in Chapter 3, the carbon price "shock" relating to policy uncertainty was taken to be symmetrical (*i.e.* the chances of the price being higher than the central expected value was equal to the chances of it being lower). This represents a situation where the policy is already in place or where the effects of uncertainty for a proposed new policy need to be analysed.

In reality however, power companies in most countries operate in a regulatory environment where the effective price of carbon is currently zero. The key risks for companies in regards to climate policy then are those that surround questions about when and to what extent CO_2 emissions will become regulated in the future. This is different from the risks modelled in Chapter 3, but a similar type of approach for analysing this problem can be conceived, as shown in Figure 30.

FIGURE 30

Framework for analysing the effects of policy timing uncertainty

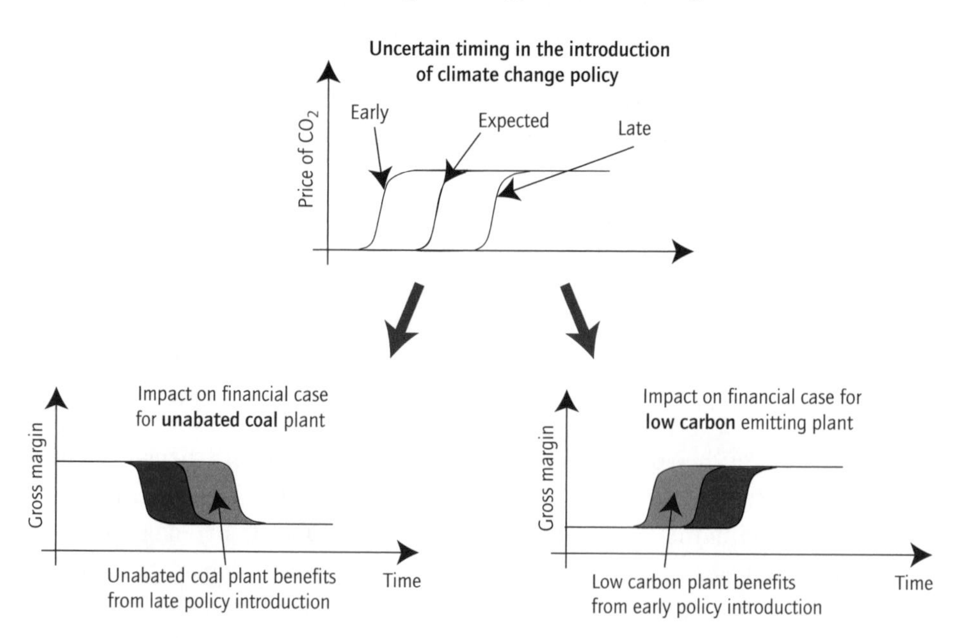

Uncertain timing of climate policy could be modelled by estimating a best guess (expectation value) of when the policy might be introduced, and then estimating a range for the earliest and latest likely introduction dates. Imagine that if climate policy is introduced at the expected time, the economic case over the plant lifetime for the two projects is the same. However, the risks will be opposite:

an earlier than expected introduction of climate policy would be beneficial to a low-carbon emitting project, whereas a later than expected introduction of policy would be beneficial for unabated coal and other more emissions-intensive plants.

This choice between two investment options that respond in opposite directions to a change in CO_2 price is analogous to the situation we modelled in Chapter 3 in the choice between a coal and gas plant. In that situation, we noted that the return on investment would need to exceed some threshold in order to overcome the value of waiting, and that this would likely result in increases in electricity prices.

In the case of uncertain policy timing, there will be an option value to waiting if waiting yields better information about when policy will be introduced. It is quite likely that waiting could yield useful information. For example, it might allow the tracking of new laws through the legislative process, and allow companies to accept and manage the results of government review processes. There may also be an inherent value to waiting simply by virtue of the passage of time once the initial estimate of the earliest introduction date has been passed. In that case, each year that passes without introduction of a policy would reduce the range between earliest and latest introduction dates.

The value of waiting could be reduced by sending clearer signals to companies about the likely timescales for policy introduction. This suggests that in situations where there is currently no stated climate change policy, the introduction of timetables and targets for future policies will help reduce the risks for companies planning investments.

Technical and other risks

The analysis presented in this study has only looked at a subset of the risks, the focus being CO_2 price risk, with fuel price risk included for comparative purposes. Clearly, the full set of risks that face companies when making an investment decision is much broader than this. Risks will depend on the particular technology, market and country being considered for investment. Although these other risks will have different characteristics from the fuel price and CO_2 price risks studied here, we can draw some general lessons from the analysis that might be useful when thinking about their impact.

We saw from the analysis that the investment threshold arises from an option value of waiting. This occurs when new information is expected to become available that would improve the investment decision being made (*e.g.* the build/no build decision or the choice of technology). In order for this information to be worth waiting for, it needs to become available independently of the investment decision (*i.e.* the information would need to arise from an event external to the project itself). In the case of CO_2 price and fuel price risk, these were considered to be exogenous variables unaffected by the investment decision.

Technical risks have a rather different nature, and are quite varied. Although all projects will have some technical risk, it is more important for less established and/or more complex technologies. In the power generation context, technical risk might be considered a significant factor for nuclear, carbon capture and storage and novel forms of renewable energy. Technical risks may include:

- Uncertainty over capital costs.
- Uncertainty over performance of a plant (efficiency, output, reliability).
- Uncertainty over operating, maintenance and decommissioning costs.

Capital cost uncertainty may be important if the technology is relatively new, but construction costs can also change even for established technologies because of changes in commodity prices. The supply of generation equipment has its own market dynamics, and prices will go up as demand increases.

Other factors that can affect project costs and/or performance include regulatory factors such as safety and planning and other local conditions placed on operators of the plant, which means that cost estimates based on experiences of building similar plants elsewhere may not accurately translate across to the new project. Technical risks can therefore broadly be split into those which are generic to the technology as a whole, and those that relate to the specific application of the technology as defined by the project, taking into account site-specific issues.

In general, technical risk is resolved through gaining experience in the application of the new technologies, and by feeding back the learning from these experiences into further development of the technology. The extent to which there is value in waiting for this type of new information depends on whether the risks are mainly project-specific (in which case waiting will reveal little or no new information), or if the risks are mainly generic, in which case new information may be revealed as a result of the experiences of other early adopters of the technology. Clearly, if everybody waits, there would never be any new information, so there has to be some mechanism by which the technology can start on its way down the development track in order for learning to take place.

Another difference is that technical risk is often dealt with as asymmetric - the new technology is being considered because it is expected to have certain beneficial characteristics, but because its performance and/or costs are uncertain, contingency costs and/or higher discount rates are used to reflect these risks in the project appraisal.

In general, a price-based policy would give rise to carbon prices that are uncorrelated with technical risks for any given project, so that the total project risk would incorporate both types of risk (the total variance in project returns being the sum of the variance caused by carbon price variations and the variance caused by technical risk). It might be possible, however, to design policies that are better matched to the technical risks that need to be addressed by having payments that in some way are better correlated with the risks being faced. This could be an interesting and fruitful area for further work.

Other important drivers of investment decisions that have been excluded from our quantitative analysis are non-climate related environmental regulations. Key examples include the Large Combustion Plant Directive in the EU and the Clean Air Act (and the subsequent Clear Air Interstate Rule) in the United States (USA) that regulate emissions of sulphur dioxide and nitrogen oxides. Regulation of other pollutants such as mercury and particulates can also play a significant role in technology choice.

Such regulations play a particularly important role for decisions relating to existing coal plants. Whilst an existing plant may be brought into compliance with the regulations by retrofitting clean-up technology, this is expensive. These retrofit investments do not make sense if the plant is relatively old, partly because the costs will be higher for an inefficient plant, and partly because there may not be sufficient time left to recoup the environmental investment costs if the plant has a limited remaining lifetime. The need to comply with these environmental regulations can therefore be an important driver in deciding on the timing of closure and replacement of older coal plants.

These environmental regulations can also interact with climate policy risks. In the USA where there is currently a zero price for carbon, climate policy uncertainty imposes an asymmetrical downside risk for coal plants, which can reinforce the risks posed by these other environmental regulations. This cumulative pressure will tend to encourage companies to consider alternative fuels and/or more advanced coal-fired technologies where the subsequent costs of environmental clean-up and carbon capture and storage may be lower. The technical risks of these newer technologies may be higher, so there is some trade-off to be made by companies in balancing the risks of the different options available to them.

Managing risk

Companies manage risk in different ways. There is usually some balance between a formal inclusion of risk in the financial appraisal of a project and a higher-level assessment, which takes account of company-wide strategic aims. Formal analysis of risk may include use of different discount rates, different internal rate of return requirements for different project types and different technologies, or more detailed "Monte Carlo" simulations, which take account of expected ranges of various uncertain variables affecting the cash flows for the project. Strategic considerations might include an assessment of whether the investment allows entry into a new market or consolidates a position in existing markets, and whether the investment fits the company's overall plans for its portfolio of generation plants.

Long-term contracts

In principle, the financial risk surrounding investment decisions could be fully hedged by selling or buying forward contracts for the electricity off-take or fuel inputs. The most direct example of this occurs in tolling agreements, where one company builds and operates a new generation plant (usually a CCGT plant), and another company contracts to supply that plant with gas and to take all the electricity. The company that builds and operates the generation plant is paid a rate (toll), which allows recovery of investment costs, and acts essentially in the same way as a leasing agreement for availability of the plant.

Although tolling arrangements are not a dominant model, they do have a role to play because of the way they divide market and technical risks amongst the two parties. Independent power producers that have a particular expertise in building and operating plants may be well placed to take on the technical and construction risks of a new plant. For independent power producers, this arrangement can be less risky than the full merchant power generation model where companies take the full range of technical and market risks. Merchant power generation investments occurred quite substantially in the United Kingdom (GBR) and USA markets in the 1990s, with substantial losses being incurred as a result of market risk in subsequent years when electricity prices fell.

The counter-parties to tolling arrangements are likely to be large integrated and/or diversified power companies who are in a better position to manage the market risks. It may suit them to have some portion of their generation plant portfolio where the technical risk is offloaded to a third party. This can have accounting benefits for the off-taking company, since there is no capital outlay required during the period of plant building. Under a tolling arrangement in the EU emissions trading scheme, carbon emissions would be allocated to the owner and/or operator, but responsibility for compliance would rest with the off-taker. Essentially climate policy risk would be taken together with other market risks by the off-taking company. An example of this type of arrangement includes the completion in 2004 of an 800 MW capacity CCGT plant at Spalding in the United Kingdom by independent power producer InterGen, which has a tolling arrangement over 18 years with Centrica, a vertically integrated generator and supplier of gas and electricity.

The degree to which technical and market risks are separated in these types of agreements vary – there is rarely a pure separation. Long-term contracts between generators and electricity users are another way in which market risks can be managed. These can be good for a plant whose cost base is expected to be relatively constant (*e.g.* coal, nuclear and renewables), and may suit electricity users who want to stabilise their electricity costs in a volatile electricity market. Such long-term contracts tend to occur in bi-lateral deals, an example being the contracts that underpin the new Finnish nuclear power station, which will be 57% privately owned, with these owners (mostly heavy industry with large base-load requirements) effectively contracted to the other owners over the lifetime of the plant. Involvement by energy-intensive industries in dedicated power generation projects may be of growing importance (Reinaud, 2006).

In principle, forward markets for electricity and fuels could play a similar role, but these markets tend not to be long enough (more than ten years) to offset risks over the lifetime of the project.[10] Instead, they are used to manage risk over a shorter time horizon. Forward gas markets in the United Kingdom typically go out to around three years, and some deals may be done based on these markets, which stretch to five years (Leppard, 2005). Electricity futures and forward contracts in the United Kingdom and Germany typically go out one to two years. In the Nordic market, forward contracts for electricity go out up to three years ahead. These forward contracts are mainly used to manage operational risk, and help companies plan their operations to balance supply and demand, with one year ahead being a typical planning horizon for many companies. The reason the Nordic market may be slightly longer is that there is a heavy reliance on hydro power, and water levels may vary over periods longer than one to two years.

Company structures and portfolios

Another way in which firms tend to hedge their operational risks to balance supply and demand is to aggregate along vertically integrated lines. Historically, in non-liberalised markets, this type of vertical integration has been the usual model in the development of the power generation industry. In liberalised markets where electricity companies were originally unbundled as part of the liberalisation process, generation companies have tended to recombine with supply companies as this can help maintain a certain level of sales for the electricity.

Clearly, if there is full competition at the retail end of the market, and if consumers are able to change their supplier without cost, such arrangements do not give guaranteed price for the electricity, since retailers need to remain competitive on price. Nevertheless, given that there is usually some inertia amongst consumers (particularly for small-scale consumers and households) vertical integration does give some protection from short-term volatility of prices. This is evidenced by the financial security of the vertically integrated companies compared to independent power producers who suffered from price risk exposure in both the GBR and USA markets in the 1990s. The presence of risk and

10. The forward market is the over-the-counter financial market in contracts for future delivery, so called "forward contracts". Forward contracts are personalised between parties. The forward market is a general term used to describe the informal market for which these contracts are made. Standardised forward contracts are called futures contracts and traded on a futures exchange.

uncertainty may also favour large companies who are able to absorb the market risks associated with the large capital-intensive projects inherent in the power sector.

Considerations of risk will often drive companies' strategies regarding the technology base for their generation. Some companies may choose to specialise in particular technologies such as renewables and nuclear if they have a competitive advantage relating to a concentration of technical expertise in these more technically risky projects. Other companies may focus on a particular technology for other reasons. Oil and gas companies, for example, tend to focus on CCGT plants as it allows them to hedge some of the price risk of gas by playing in both electricity markets and gas markets. Moreover, being a supplier of gas as well as a gas-fired generator may give the company a competitive advantage in this technology.

A considerable number of generators, however, will favour some kind of generation portfolio with a mix of different types of generation. These types of generators likely form the core of generation capacity in most countries. Several companies interviewed maintain guidelines for the overall portfolio mix that they wanted to achieve. Usually these are used to help guide strategic direction rather than acting as "hard" targets. In some cases, company portfolio mixes have been largely inherited from historical ownership patterns (*e.g.* pre-liberalisation) and subsequently maintained. In other cases, a mixed portfolio has arisen in response to the different investment conditions that pertained at the time when plants were being built. As investment conditions changed over time, so did the type of plant, until a mixed portfolio including coal, gas, nuclear and renewables was achieved.

Companies often see a benefit in being able to spread their exposure across several different types of risk through portfolio diversification, even though financial theory would suggest that there is no formal benefit from doing this at the company level (see Chapter 1). This is partly to do with the way in which companies interact with financial markets, which is not done according to the theoretical models. We explore the ways companies interact with financial markets in a bit more detail in the next section. The implication of this is that, in theory, companies may respond to climate policy risks by hedging through greater diversification, although the results in Chapter 3 indicate that climate policy risks would probably act as a weaker driver of diversity than fuel price risk.

Financial market responses

A detailed review of corporate and project finance is beyond the scope of this report, but it is useful to review some of the important interactions between financial markets and the power generation sector relating to the perception and management of risks.

Broadly speaking, power generation projects may be financed in one of two ways: project finance and corporate finance. Project finance refers to the case where money is raised from financial markets specifically to build and operate a new plant (sometimes through a company set up specially for this task). Corporate finance refers to the case where a project is financed directly on the company's balance sheet.

In many cases, the economic decision on whether or not to go ahead with a project is taken independently from decisions about the financing of the project. Money can be borrowed from financial markets either in the form of debt or equity (with various shades of grey in between reflecting the order of priority in which investors are paid). Debt requires repayment at a fixed rate, whereas equity payments will depend on the financial performance of the company, and is only paid once debt payments have been settled. Debt and equity stakes in projects and companies therefore attract different levels of risk and return.

Companies often raise long-term corporate debt by issuing bonds. Bonds are issued by companies promising to pay a certain percentage return each year plus repayment of the capital at the end of the period of the bond (*e.g.* 10-15 years after issue). The bonds are purchased from the company by bond traders, and re-sold in secondary bond markets. Once the bond has been issued, the annual payments made by the issuing company are fixed for the duration of the bond. However, the value of the bond in the secondary market goes up and down depending on the probability of default on bond payments, which is an indicator of the credit risk faced by the issuing company. The credit rating of a company therefore affects the amount of money that companies will receive when they are issuing their bonds in the first place – an important element in the cost of raising debt capital. Companies consequently have to demonstrate to financial markets that their planned investments and the risks to their existing assets do not overstretch them financially, thereby increasing their possibility of default. Active bond investors would assess these risks on a company-by-company basis rather than assuming some type of industry equilibrium CAPM-type model. This is partly because debt investors cannot diversify away company-specific risks in the same way as equity investors since there is no upside risk, only downside risk of default.

Although a new source of risk in the market, hypothetically, could affect the cost of debt by increasing the possibility of financial distress, bond markets have so far not reacted measurably to climate change policy risk, at least in relation to the power sector. In Europe, the free allocation of allowances in the EU emissions trading scheme has increased the asset value of most power companies, effectively reducing their credit risk. In the longer term, if free allocation is phased out, climate policy risk may become more material. Credit ratings agencies certainly include possible changes in climate policy in their analysis of credit risk in the power sector (Standard and Poor's, 2006).

Banks also provide debt to companies, and are often particularly involved in financing "special project vehicles" where finance is raised for a particular investment. Banks will do a detailed risk analysis of the project, including market, regulatory, construction, operational and contractual risks. Fundamentally, the concerns that underpin a bank's valuation of project risk are similar to those for a bond investor at the corporate level, and relate to the risk of default on loan repayments. Risk exposure tends to be higher for individual projects than for the more diversified operations at the corporate level, so that the costs of debt would, in general, be higher for project financing. There can however be a grey area between project and corporate financing; for example, if there are strong links between the banks and the corporation backing the project. In Japan, government-owned banks such as the Development Bank of Japan have a closer relationship with government ministries and therefore have a different risk exposure, giving them a greater ability to extend long-term loans to power companies for a new plant than commercial banks.

The other major determinant of the cost of capital is the cost of equity. In principle, following the CAPM model formulation discussed in Chapter 1, the cost of equity should only take account of systematic (market) risks, and ignore specific (diversifiable) risks. In practice, separating systematic and specific risks is not straightforward, particularly when assessing the effects of a new source of risk that has not yet been reflected in historical data of market returns.

Equity investors would generally value a power company based on a discounted projection of its expected future revenues. The discount rate would take account of the cost of capital given the particular capital structure (*i.e.* the debt/equity ratio) of the business. The cost of equity would take into account the risk-free rate and a view on the appropriate beta factor for the company, which might be based on *ex post* analysis of historical data.

Uncertainty about future revenues should in theory be incorporated into the assessment of the company's beta factor value, but in practice getting an objective ex ante determination of future changes in the beta factor is difficult, and usually a more pragmatic approach is taken. For example, a sensitivity analysis of the company's cash flows might be undertaken to derive best-case, worst-case and central-case scenarios, with a probability weighting given to these. Alternatively, a different assumption may be used about the cost of capital in order to reflect a change in risk level.

In general, equity investors will try to determine whether the risks faced by the business they are investing in are adequately reflected in the returns to be gained from share ownership. Often, this analysis will be carried out at the level of individual companies, and equity investors would expect to be recompensed for taking risk rather than actively seeking to neutralise the risk by spreading their bets. Therefore, to the extent that new company-level risks are seen to be material, they are quite likely to impact on the cost of equity, at least in the short term. In the longer term, additional variability in company returns would be incorporated into an ex post analysis of the company's beta factor, and so presumably, company-specific risks would feature less strongly in the calculation of the cost of capital.

In recent years, institutional investors have actively encouraged companies to disclose information on the business risks and opportunities presented by climate change. Perhaps the most high profile investor initiative has been the Carbon Disclosure Project, now supported by 211 institutional investors representing more than USD 31 trillion in assets under management. Institutional investors have also contributed to the Global Reporting Initiative (which nearly 1,000 organisations in over 60 countries currently use as the basis for their environmental and social reporting) and, in May 2005, 14 leading investors and other organisations worldwide launched a new effort to improve corporate climate change disclosures – the Climate Risk Disclosure Initiative. These initiatives - in particular the Carbon Disclosure Project - have played an important role in raising the awareness of climate change as a business issue and have contributed to companies increasing their reporting in this area, although questions remain as to the usefulness of information currently being provided as a result of these initiatives.[11] Nevertheless, power companies are now paying more attention to how their investment decisions will be perceived by financial markets than was the case even a few years ago. A European group called the Institutional Investors Group on Climate Change (IIGCC) representing assets of EUR 1.4 trillion have been vocal in encouraging the formation of clear policy guidance on climate change, described in a recent statement released by the group:

...Climate change presents a series of material business risks and opportunities - for investors and companies - to which investors must respond. Despite remaining scientific uncertainties, we believe that it is appropriate to adopt a proactive approach to this issue and to take action now that will result in substantial reductions in global greenhouse gas emissions within a timeframe that minimises the risk of serious impact ...It is our view that governments should lead this response by creating a framework that provides incentives and investment certainty to companies and individuals....

Source: Investor Statement on Climate Change, IIGCC
www.iigcc.org/docs/PDF/Public/IIGCC_InvestorStatementonClimateChange.pdf

11. www.insightinvestment.com/Documents/responsibility/climate_change_disclosureinitiatives_report.pdf

IMPLICATIONS FOR POLICY MAKERS

Climate policy uncertainty in this analysis has been expressed in terms of uncertainty over the cost of emitting carbon dioxide using carbon price as a proxy. High carbon prices are indicative of stringent climate change policies (which are assumed to cost more to achieve, whatever the type of policy used), and conversely, low carbon prices reflect less stringent climate policy. The approach therefore provides a way of modelling uncertainty effects in relation to different policy design options, including other types of policy beyond price-based policies (*e.g.* technology standards and subsidies).

The model is relevant to understanding how companies might respond either to an existing policy or to the introduction of a proposed new climate change policy, where there is some uncertainty over the duration of prices determined by this policy. The results indicate that climate policy uncertainty would lead to a value of waiting, which could lead to 5%-10% rises in electricity prices and a preference for investing in low capital-cost options including delaying plant replacement. In order to achieve the same level of investment in low-emitting technologies, climate change policies will have to be effectively more stringent in the real world of uncertainty than in an idealised case without uncertainty.

Similar considerations will apply in situations where the timing of the introduction of climate change policy is the key uncertainty. Although we have not included this in the model explicitly, insight suggests the model can be extended to this situation. The value of waiting arises from the fact that companies might choose to invest in different types of plants depending on the timing of policy, with late policy introduction favouring a higher emitting plant and early policy introduction favouring a lower emitting plant. Narrowing the possible time range for the introduction of climate policy could help to reduce the investment risks.

There are a variety of possible models and variables to project energy needs that can be used to tailor climate change policies. This book used models specifically adapted to consider ways to formulate effective policies. Effective polices are those that will not hamper the future energy demand and supply structure through creating unacceptable investment risks for firms that supply energy. Other models can be adapted to assess similar or related policy and risk uncertainty issues. In the first section of this chapter, we explore a few of these to highlight the implications of their use for policy formation.

The second section in this chapter looks at some practical considerations around creating policy certainty in different policy contexts. The discussion draws on the views expressed during the consultation phase of this work.

As economic and market conditions vary considerably across different IEA member countries, so the financial case around the decision to build one type of generation plant over another will also vary from market to market. The investment thresholds described in Chapter 3 do not imply that one particular technology is preferable to another - the underlying economic case for a particular technology may easily outweigh the investment thresholds we have presented, depending on the particular circumstances companies face. The analysis illustrates the effects uncertainty may have in addition to the underlying economic case when decisions are made. Hence, policy makers can use this type of approach in conjunction with their own analysis of the underlying economics of different generation technologies. The thresholds shown in Chapter 3 are based on simple assumptions to maintain transparency and draw out general conclusions. For a more detailed analysis, assumptions would need to be adapted to better suit country-specific conditions. This is discussed further in the third section of this chapter.

Implications for energy projections

Many different types of models are used by energy policy makers to help them understand trends in energy consumption and supply, and to tailor policies to meet particular policy objectives. These "top-down" energy models typically aim to describe the energy system at the macro-economic level, often at the national, regional or global scale. A review of different energy model structures by the Energy Modeling Forum yielded a categorisation of model types, including:

- equilibrium models, of which there are three types, namely, disaggregated economic equilibrium, aggregate economic equilibrium and energy-sector equilibrium models;
- energy-sector optimisation models; and
- energy sector regression models (Beaver, 1993).

There are important variations in these and other models. For example, simulation models aim to provide a longer-term focus by combining a backwards-looking econometric regression analysis with more judgemental parameters of future trends. Top-down models are usually deterministic in the sense that given a particular set of input parameters, they will give a unique solution to the question being posed. For example, the models might give a projection of the cost and/or optimal technology mix for the supply and consumption of energy under different constraints and market conditions. Clearly, the future value of many of the input parameters to these models will be uncertain. This uncertainty is usually dealt with through scenario analysis, which may include a risk-adjusted discount rate.

However, two problems arise from these approaches. First, with scenario analysis, there is usually no probability weighting given to the different scenarios. The model results give a sense of the possible outcomes, but without some idea of the probability of these outcomes, it is difficult to get a sense of the risk associated with different policy choices. This problem may, to some extent, be overcome using more systematic approaches, such as testing the range of model outcomes arising from feasible ranges of input parameters using techniques such as a Monte Carlo analysis or Latin hypercube sampling. These rely on multiple model runs and take better account of the interactions between the different uncertain parameters. Second, simply using a risk-adjusted discount rate may not be adequate, as it assumes that uncertainty is resolved in a smooth linear fashion, and does not account for adaptive behaviour. As described by Kann and Weyant (2000; 37):

Optimisation models search for the most efficient way to address a problem, given our current state of knowledge. The distribution that results from propagating uncertainty through an optimisation model thus needs to be interpreted as follows: each point on the output distribution represents the result of an optimisation for a particular uncertain state of the world represented in the input variables. This implies a "learn now then act" approach in which the uncertain state is revealed before action is taken...In reality, the policy-maker's situation is not one of resolving uncertainty now and acting optimally according to the revealed knowledge...Knowing that policy X is the best policy in the state of the world x and that policy Y is the best policy in the state of the world y does not tell us how to choose an optimal policy before the state of the world is revealed...A more appropriate approach of including uncertainty in optimisation models would be to create a sequential decision making model.

There is a body of literature that explains the theory of why different results will be obtained by putting uncertainty "inside" the scenario rather than allowing the model to behave as though it had perfect foresight within any given scenario (Bunn, 1986; Louveaux and Smeers, 2005). These insights result from a basic economic insight known as Jensen's Inequality. This states that if $F(x)$ is a function of an uncertain variable x, then the expected value of that function $E[F(x)]$ will generally be different from $F(E[x])$, the value that the function takes when it is evaluated for the expected value of x. We can express this mathematically as:

$$F(E[x]) \neq E[F(x)]$$

This is because in general the function $F(x)$ is non-linear, so that simply taking the expected value of x and evaluating the function for this single value does not replicate the a mapping of the uncertain values of x onto equivalent uncertain outcomes in the value of the function. The strength of this inequality depends on the non-linearities involved.

The analysis presented in this book has been dealing with non-linearities introduced by the assumed flexibility of managers to optimally time their investments in relation to uncertain events. We have shown that in some cases, this flexibility (*i.e.* the option to wait) can be quite valuable, reflecting the ability to avoid downside responses and maximise upside responses to price uncertainties. This ability to adapt behaviour leads to an asymmetrical profit expectation, even if the initial price risk is symmetrical, and is therefore an important non-linearity of which to take account.

We can see this logically by thinking about how the choice between a gas and a coal plant under carbon price uncertainty might be modelled. A typical scenario approach might be to identify two different carbon price paths over time, representing credible estimates of upper and lower bounds. A deterministic model would then look at the economics over the lifetime of the plant including the (deterministic) carbon prices, and provide an optimal decision for each price scenario. We might imagine that under a low-carbon price scenario the model decides to build a coal plant, and under a high-carbon price scenario the model decides to build a gas plant. Allowing for sequential decision making in response to uncertainty gives a different result, along the lines we have seen in Chapter 3. Faced with uncertainty over whether carbon prices will follow a high path or a low path, companies will exercise their option to wait unless electricity prices rise sufficiently to compensate for the risk of making the wrong decision. This is a different result, which would not have been obtained using a deterministic, scenario-based approach.

The "first-best" solution to this problem would be to adapt top-down models to allow for sequential decision making. The difficulties involved in doing this are described in Kann and Weyant (2000). Optimisation models tend to assume that optimal decisions and policies are determined only once at the beginning of the run based on currently available knowledge. They would therefore require structural changes in order to incorporate multi-stage uncertainty. The size and complexity of many top-down models may be prohibitive to this. A partial solution may be achieved by allowing for a two-stage model where the first stage consists of decisions taken before the uncertainty is resolved, and second stage decisions are taken after the uncertainty has been resolved. The set of second stage decisions can be different depending on the outcome of the first stage. A review of two-stage stochastic linear programmes is given in Birge (1997).

An alternative or "second-best" solution would be to run a separate stochastic optimisation model in parallel with the top-down macro-economic model. The effects of uncertainty could then be generated off-line, in a manner similar to that described in this book, but with a set of assumptions that are consistent with the macroeconomic model.

The investment thresholds generated through such a stochastic optimisation model could then, in principle, be incorporated into the macro-economic model. There are a number of different ways in which this could be done; the most appropriate method would probably depend on the structure of the top-down model in question. For example, time-dependent investment thresholds of the sort presented in this book could be incorporated into the technology cost assumptions, or alternatively expressed in terms of a cost of capital and introduced to the model through a revised discount rate (again, this might have to be time dependent).

We have made the case in here that the effects of uncertainty are, to a first approximation, independent of the economic assumptions about the investment being modelled. This independence would allow the results of an off-line calculation of risk-premiums to be transferred across reasonably well into a macro-economic model. There are limits to this independence, and it may be important to take account of circumstances where the uncertainty is linked to the underlying economic assumptions. It seems likely however that these circumstances could be adequately accounted for, provided that careful consideration was given to the design of the stochastic optimisation model and the way in which the results were incorporated into the macroeconomic model. This second-best approach could provide considerable benefits in helping understand the likely evolution of the energy sector under uncertainty, whilst perhaps being significantly more practicable than the first-best approach of major restructuring of top-down models.

Creating policy certainty

Policy duration

The discussion in this book has been about climate change policy risk from the perspective of electricity generation companies. It is also valid to look at the same problem from the perspective of policy makers. The needs of investors in regards to long-term policy should be weighed against the benefits of policy flexibility. A flexible policy responds to improved information on climate change science, estimates of the costs and benefits of mitigation, as well as political decisions and trends in other countries. The trade-off between flexibility and certainty will depend on the total cost to companies for dealing with the risks posed by policy uncertainty compared to the benefits to the economy as a whole of maintaining a flexible policy position in order to be able to respond to increased knowledge about optimal mitigation responses to climate change.

This trade-off will depend on who is best placed to hold the risks associated with uncertainty over climate change policy impacts. Although companies are in a strong position to manage risk generally, climate change policy risks are different in nature. These risks are long-term, political and dependent on the outcome of international negotiations and agreements. In particular, company decisions depend on whether or not climate change policies take account of complex economy-wide decisions about balancing uncertain mitigation costs against uncertain adaptation costs. It is arguable that in this context, governments are the only ones capable of underwriting such long-term risks. Such a situation would not be unique to climate change policy risks – governments have typically needed to maintain responsibility for long-term liabilities associated with nuclear power plants, for example.

There is however a downside for companies if policies are set too far ahead. When considering what targets to set, governments would need to take account of the asymmetric risks of climate change impacts (*i.e.* the possibility of "nasty surprises"). The longer the period of time over which a policy is fixed, the less flexibility the government has to respond to new information. Very long-term targets would therefore need to be more stringent than shorter-term targets in order to insure against worse-than-expected impacts of climate change.

On the other hand, we have shown in Chapter 3 that policies with a shorter duration need to be more stringent in order to stimulate investment in low-carbon technologies because of the additional investment risk. This implies that for a given cost level, there should be a compromise between policies that minimise overall risk and policies that maximise environmental benefits.

The critical period for investors building a new power plant is the period 5-15 years into the future, as this is the period when the plant will be up and running and contributing most to the discounted cash flow. There are diminishing returns in setting targets beyond this period, since the discounted present value of annual returns that far in the future become significantly less important.

Given the multi-decadal timescale of many of the worst climate change impacts, it seems unlikely that committing to policy targets 10-15 years ahead would seriously constrain governments' abilities to manage climate change impact risks. Although in order to get a better understanding of where an appropriate balance might lie, further work would be needed to estimate the overall economic benefits of policy flexibility.

Creating certainty in emissions trading schemes

Emissions trading schemes are complex instruments with many details needing to be specified as part of the policy, such as:

- The type of system to be implemented (*e.g.* cap and trade, baseline and credit, absolute versus intensity targets).
- The scope of the scheme–including the sectors to be covered, the definition of installations to be covered including any size restrictions, and which greenhouses gases are to be included.
- The overall cap or intensity targets.

- Allocation methodology, particularly the level of free allocation of allowances/credits.
- The treatment of new entrants and plant closures.
- Definition of monitoring and verification rules, enforcement and penalty regimes.
- Recognition of JI and CDM credits and links with other emissions trading schemes.
- Administration of the scheme, and procedures for review of the scheme rules.

Even if climate policy targets are set for a 10-15 year period, policy details are likely to need to evolve and adapt in order to learn from experiences of their implementation. This is particularly true during these early stages of policy development. The scope of the scheme in the early stages may be deliberately limited in order to build experience with the simplest power generation sectors before adding greater complexities.

Most companies identified the likely supply of allowances and/or credits in the market as the most important issue for creating long-term certainty. Perhaps the most important element of this is the overall cap set for the scheme, and how far into the future this is set. However, the supply of allowances will also be strongly affected by the rules on the use of JI and CDM credits. This implies that other fundamental changes to an emissions trading scheme such as an expansion of the scope, structural changes (*e.g.* to the type of target or penalty regime) and links to other schemes would have to be managed carefully to keep the expected balance of supply and demand within bounds similar to those prior to the change.

Companies considered the level of free allocation as a secondary concern, although fair treatment was seen as very important. A general principle identified by many companies was to make policy decision making as transparent as possible. Companies are generally confident in committing capital to projects, even in an uncertain environment, as long as they can establish a competitive advantage in the market. When it comes to regulatory risk, this requires that policy makers establish clear rules, and that companies can be confident that these rules will be applied consistently to all market players.

The more complex the policy, and the greater the scope for exemptions and special treatment, the less certainty there will be for companies. Several companies privately agreed that auctioning of allowances–rather than free allocation based on historical emissions–would provide greater transparency. In particular, auctioning allows a much more transparent treatment of new entrants, and reduces distortions around the timing of plant closures.

Given the likelihood for policy changes to alter the price of carbon, several general commentators have called for more direct controls on price by using taxes and trading instruments. For instance, Pizer (2002) and IEA (2002) suggested that a hybrid approach that combines elements of quantity and price would be the most effective. This could be done by introducing a price cap and a price floor to an emissions trading market that would essentially limit the variability in carbon price. The tighter the range between the floor and the cap, the closer the policy would come to a tax, while the wider the range, the closer the policy would come to a pure emissions trading scheme.

As we saw in the analysis in Chapter 3, price caps would reduce investment thresholds for an emissions intensive plant, and price floors would reduce investment thresholds for a low emitting plant. An approximately symmetrical arrangement of a price cap and floor around a central expected carbon price would not distort the economic case for either choice, but would bring down overall investment thresholds. Price caps may be politically attractive. It has been argued by Philibert (2006) that ensuring an upper limit to price variation may buy a greater level of political commitment to emission reductions. It could increase the willingness of countries to participate in international emissions trading schemes, and might facilitate the adoption of more ambitious policies that would result in higher expected environmental benefits.

Conversely, given the tendency of companies to delay plant replacement in the face of policy risk, there is perhaps an even stronger argument in favour of price floors to support investment in low carbon technologies in order to assist the transition towards a near-zero emitting energy infrastructure. Various proposals have been made for how price floors could be introduced. Helm and Hepburn (2005) have proposed a carbon contract arrangement by which governments could contract ahead with companies for the supply of emission reductions over long time periods to back up their political commitment to long-term targets. From the companies' point of view, such long-term contracts would provide a guaranteed source of income for their low carbon investments. Helm proposes to auction these contracts; ensuring long-term emission reductions were achieved at the lowest possible cost.

Ismer and Neuhoff (2006) extend these ideas to define a carbon option contract, which gives companies the right but not the obligation to sell allowances to the government at a certain price. Under this arrangement, companies investing in low carbon technologies would not carry the risk that the market price might turn out higher than the strike price in the contract. Governments, on the other hand, would be fully committed to a floor price, creating an incentive for them to maintain sufficient policy stringency to keep market prices above the strike price in order to manage their liabilities.

The results in Chapter 3 indicate that the key to the success of such schemes would be the ability to ensure the credibility of the established price over long periods into the future. The contractual nature of these schemes may provide sufficient legal basis to provide such credibility.

Hepburn *et al.* (2006) show how price support could be achieved in the context of Phase II of the ETS (2008-2012) by auctioning 10% of allowances with the use of a minimum reserve price. This would not guarantee a floor price if the market price were considerably below the reserve price. However, in this case, the allowances would then be withheld from the market, which would at least partially support the price.

Non-price based policies

Different policy types present different risks to power companies. However, a full analysis should take account of the extent to which policy risks will be incurred outside of the power sector, including electricity consumers. We can illustrate some of the main considerations with some simple "thought experiments".

Non-price-based policies could be designed in a way that, in principle, creates very little risk to a power generation company; but this will also depend on the type of market in which the generator operates. For example, imagine a mandatory technology standard that requires the use of carbon capture and storage for all new coal plants. If this is imposed in a market with price regulation, the investment and operating costs of the new technology should be recoverable through a revised electricity rate agreement with the regulator. Assuming the appropriate rate can be negotiated, the power generation company would face very little risk. If the climate policy subsequently collapses (*e.g.* dropping the mandatory requirement for CCS), then the sunk costs of previous investments would in principle still be recovered through the agreed rates.

However, the risk has not just disappeared. In the event of a collapse in the climate policy, the electricity users would continue to pay for the sunk costs of the CCS that had already been invested in prior to the collapse. It is therefore the electricity users in this case who effectively take on the policy risk,[12] although because the risk would be widely distributed, it is less visible than when it is concentrated in a few generation companies.

In a competitively priced market, a technology standard that required the adoption of CCS would need to be implemented via a price support mechanism, such as a feed-in tariff or an obligation for the supply of a certain level of generation from that technology in order to bridge the cost gap between CCS and unabated coal. This effectively replaces the carbon price support with a different type of support mechanism. The model results in Chapter 3 regarding the effect of policy duration and price controls would also apply to this type of policy approach.

An alternative policy would be a direct subsidy to promote early adoption of new technologies, which might be particularly suited to addressing technical risks. Subsidies can take various forms including direct capital grants and favourable tax treatments. Imagine that a policy is introduced to subsidise a number *(n)* of new plants in order to improve information on the performance and costs of a new technology. These subsidies could be designed to offset the technical risks for these plants. Nevertheless, investment in these new plants does not create a new market for that technology (unless *n* is very large and brings the costs down below existing technology). Companies are unlikely to invest large amounts of their own time and research and development budgets to the problem of technology development and deployment unless they see a strategic reason for doing so.

Although direct subsidies, therefore, could minimise the investment risks associated specifically with the these plants themselves, the full benefits of learning, information acquisition and overall reduction of technical risk for the technology more generally will not be gained unless there is a reasonable prospect of longer-

12. The interests of electricity users in a price-regulated market are represented in the negotiation between the regulator and the power generator, so that in practice, the power generator would not be able to pass on all the risk.

term, wide-spread deployment of the technology from which the companies could profit. New markets need to be created by setting the appropriate framework for companies to profit from investing in the new, low-emitting technologies, and allowing them to capture the learning associated with these investments.

International action

Many companies made it clear during the consultation stage of this work that an important element of policy certainty is the credibility of policy objectives. Creating policy credibility will require more than changes to domestic policy design. What really underpins the credibility of long-term domestic policy is a sense of consistency with overall international action. Firstly, consistency with international action is needed to overcome resistance relating to competitive distortions. Secondly, international action is vital in terms of creating a sense of the market "fundamentals" that underpin new investment, both in the narrow sense of a carbon price established through international credits, and more strategically in terms of a sense that there will be important new emerging markets for low carbon technologies around the world.

Given the need for international co-operation to tackle climate change, there is probably no quick fix for creating policy credibility at the domestic level. Rather, credibility will result from the accumulation over time of the experiences and actions taken by many different players around the world—both governments and companies. Each individual climate change initiative will only have limited credibility in its own right, but will add to this accumulation process.

This raises an important question about the value of linking different international climate policy systems together. At the level of the individual investor, creating such links could increase the complexity of the system, making it more unpredictable. For example, allowing JI and CDM credits to be used in emissions trading schemes means that the price of carbon in those schemes is not just subject to the supply and demand and marginal cost of abatement for the players within the scheme, but is also subject to the availability and cost of eligible projects in the host countries of the JI and CDM projects. With more "moving parts" the cost of meeting emission caps is subject to many more variables and uncertainties.

On the other hand, many large emitting companies will have commercial interests spanning a wider geographical region than most domestic policies cover, and will therefore have a strategic interest in linking up different regulatory regimes, even if this appears to lead to greater complexity. Furthermore, with several regions linked together, it makes it less likely that actions by any one country could lead to completely undermining the fundamental market value of reducing carbon emissions. The JI and CDM initiatives are seen as crucial steps in getting transition and developing economies integrated into the global framework for action on climate change. The strategic benefits probably far exceed the "local" difficulties that the additional complexity causes.

Future directions

New technology choices in fossil-energy-fired and nuclear power generation, the focus of this book, are only one part of the strategy to abate future emissions from power generation. End-use energy efficiency and renewable energy are other critical components, as illustrated in recent IEA publications such as *Light's Labour's Lost* (IEA, 2006a), *Energy Technology Perspectives* (IEA, 2006b), *Renewable Energy - Market and Policy Trends in IEA Countries* (IEA, 2004) and the *World Energy Outlook* (IEA, 2006c). The IEA's publications in energy efficiency show a significant potential for improved energy efficiency at the end use which could bring significant CO_2 reductions at no net cost to society. The IEA's research in renewable energy shows that some grid connected renewable energy technologies, such as wind power technology, has passed through the research, development and deployment (RD&D) stage and been used in power generation. There is an urgent need to do further research on how government climate change policies will facilitate the investment and deployment of energy efficient technologies and renewable power.

Besides the issues of energy efficiency at the end-use and renewable power technologies, technology risks also significantly affect the decisions of the power investors. Plant investment costs, the time period of construction, the quality of equipment and the know-how of the operational personnel all pose technical risks to power generation. There is a need to establish the best way to incorporate the technical risks into financial risks for the different power generation technologies.

Finally, both government policy makers and chief executive officers of power companies would like to know the impact of overall government climate change policies (including, carbon tax, renewable power green certificates, energy efficiency rebates,and similar schemes) on power companies' investment portfolios. These include various power generation technologies, power transmission and end-use distribution. The IEA plans to address all the above issues in the next stage of this on-going project.

Led by Dr. Richard Bradley of the IEA, this policy uncertainty analysis project began in 2005 and will last for four years. Over the past two years, this project focused on model and methodology development, database development and case studies for fossil-fuel fired and nuclear power technologies. So far, the deliverables of this project include an information paper (IEA, 2006d), a working paper (IEA, 2007) and this book. In the next two years, the project team will focus on the four issues stated above: energy efficiency technology investment versus power generation investment; technology risk assessment; government policy impact on wind power development; and overall government climate change policy impacts on power company's investment portfolios. In 2007, the IEA will deliver at least two papers for the Standing Group on Long-Term Co-Operation (SLT) of the IEA and for the Thirteenth Conference of Parties (COP13) of the United Nations Framework Convention on Climate Change (UNFCCC) where we will present our findings for the first three issues mentioned above. In 2008, the IEA will publish another book that will address the last issue.

Concluding remarks

In this book we have blended quantitative and qualitative methods to ascertain energy investors' risks caused by policy uncertainty, *and* to ascertain possible effective policy directions to lessen risks on both sides. The quantitative analysis–performed by substituting carbon prices for policy uncertainty risks using real options models–provided a mechanism to measure energy suppliers' risks. The qualitative analysis–gathered through researching the opinions, attitudes and experiences of energy suppliers' assessment of risk–gave us examples from the real world. Taken together, they provide valuable insight for policy makers to use when designing climate change policies.

The dilemma, however, is how to ensure policies are effective for seemingly different issues: those of supplying energy while making a profit and those of curtailing climate change. Looking at this challenge alongside current policies and creating future ones, we found several promising options that can take into account the quintessential triumvirate of needs: the needs of energy users, the needs of energy suppliers, and the needs of policy makers. Working together to accomplish one singular need–that of protecting the global environment– is the way forward.

APPENDIX 1 - TECHNOLOGY ASSUMPTIONS AND RESULTS SUMMARY

TABLE 5

Technical assumptions used in the model

Project specific assumptions	New coal	Retrofit CCS to coal	New gas	Retrofit CCS to gas	New nuclear	Re-powering existing coal plant		Existing oil plant
						Existing coal	CCGT	
Project lifetime (years)	40	40	25	25	40	25	25	11
Capacity retrofitted (MWe)	1,350	1,086[1]	1,350	1,2081	1,350	472	472	500
Capital "overnight" cost (USD/kW)	1,320	810	589	430	2,528[2]	NA	631	(140)[4]
Construction period (years)	3	2	2	2	6	NA	3	0[5]
Capacity/load factor	85%	85%	85%	85%	85%	85%	85%	40%
Average generation efficiency	46%	37%	57%	51%	-[3]	33%	58%	30%
CO_2 emissions factor for fuel (tCO_2/TJ input energy)	95	95	56	56	0	95	56	77
Fixed O&M (USD/kW-Yr)	42.5	65	42.5	65	992	30	20	12
Variable O&M (USD/MWh)	0.0	7.4	0.0	3.54	5.2	3.0	2.0	3.3
CO_2 abatement factor (CCS)	-	86%	-	86%	-	-	-	-

Notes: (1) This represents a loss in output from the plant due to the power requirements for the CCS, which reduces the power available for revenue; (2) Capital and fixed O&M costs are increased by USD 840/kW and USD 32/kW-yr representing 2 standard deviations above the average in NEA/IEA 2005; (3) Fuel cycle costs for nuclear are included in the variable operating and maintenance costs; (4) This is the costs for decommissioning the existing oil plant; and (5) Decommissioning may take some time, but it is assumed that once the decision is taken, the fixed and variable maintenance costs could be immediately stopped.

Table 6

Summary of investment threshold results

Investment thresholds USD /kW	Gas on margin			Coal on margin		
	CO_2 uncertain only	Fuel uncertain only	CO_2 and fuel uncertain	CO_2 uncertain only	Fuel uncertain only	CO_2 and fuel uncertain
Coal	56	320	318	58	87	170
Gas	0	113	129	291	259	306
New nuclear	78	264	322	322	57	354
Coal vs. gas choice	135					
Replace existing coal with gas	277					
Retrofit CCS to coal	236					
Retrofit CCS to gas	149					

Note: the CO_2 uncertain case includes a price jump after 10 years. CO_2 prices also have an annual random walk of ±7.75%. Fuel price uncertainty comprises an annual random walk of 7.75% for gas price and 1.8% for coal price.

APPENDIX 2
TECHNOLOGY INTERACTION EFFECTS

Chapter 3 presented the investment thresholds for coal, gas and nuclear options. Those investment thresholds were calculated for the three individual technology options considered in isolation, and did not include any technology interaction effects. These can result in an additional value of waiting arising from the fact that these different investment options can sometimes act as a mutual hedge because their profitability responds in different directions to changes in prices. This effect was seen to be important in the context of a coal versus a gas plant choice when only CO_2 prices were uncertain. The purpose of Appendix 2 is to investigate the importance of this effect in the case of the three-way decision between coal, gas and nuclear. Specifically, we investigate the following questions:

- Does the inclusion of nuclear further increase the technology interaction effects compared to the two-way coal versus gas choice in response to CO_2 price uncertainty?
- How does the inclusion of fuel price uncertainty in addition to CO_2 price uncertainty affect the technology interaction effects for the two-way coal versus gas choice and the three-way coal, gas and/or nuclear choice?

In order to answer these questions, we need a different way to represent the results, as the actual threshold values are not always straightforward to derive when there are multiple technologies and multiple stochastic variables. Instead, we present the results in terms of the option value of the investment option. Although this does not give a direct value for the investment threshold, it does allow an assessment of the extent to which adding additional technologies to the mix changes the risk profile of the investment, and also allows analysis of the relative importance of fuel price and carbon price uncertainty when dealing with a portfolio of investment choices.

Figure 31 shows a schematic diagram of the value of an investment option as a function of the expected net revenue arising from the project. This is another way of visualising the option value arising from investment flexibility in the face of uncertain revenues. The straight blue line is simply the expected net revenue minus the investment cost, and represents how that investment option might be valued under deterministic conditions. In the presence of uncertainty, the option to invest gains additional value (the red curved line) because of the additional value embodied in the ability to optimise the timing of investments with respect to the uncertain revenues. When the value of holding the option (red curve) exceeds the expected value of the project revenue (straight blue line), the option would not be exercised (*i.e.* the investment would not go ahead). Only when the two lines converge (*i.e.* when expected project revenues exceed the investment threshold) would it be optimal to invest in the project itself, rather than holding on to the option.

The results in this Appendix are presented in terms of the extent to which the option value exceeds the expected project revenue (*i.e.* the gap between the red curve and the blue line).

Figure 31

Relationship between option value and investment threshold

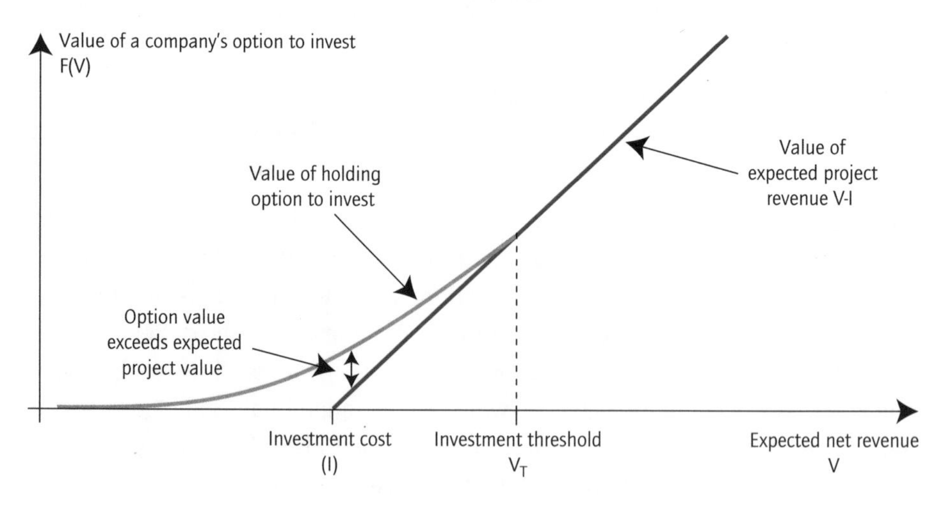

Investment would not proceed if the value of holding the option is greater than the value of the expected project revenue. This occurs when the expected project revenue is below the threshold.

The option values for coal, gas and nuclear investments (and combinations of these) are shown in Figure 32 and Figure 33 for different marginal plant assumptions. These show option values for different combinations of investment opportunities. The columns where there is only one technology named represent opportunities for investment in those technologies only. Where there are multiple technologies, we are allowing the model to make an optimal choice between the two or three named technologies.

The y-axis in these figures is the same as the y-axis in the schematic diagram (Figure 31). The lower (blue) section of the bars is the normal NPV that would be calculated for the different projects under deterministic conditions (*i.e.* the same as the blue line in Figure 31). Initial conditions have been deliberately chosen to make these approximately equal for the three technologies. The middle (red) section of the bars is the additional option value that arises as a result of the flexibility to optimise the timing of investment in the face of uncertain prices. The upper (beige) section of the bars is the further option value that accrues when considering two or

FIGURE 32

Option values when there is variable plant on the margin

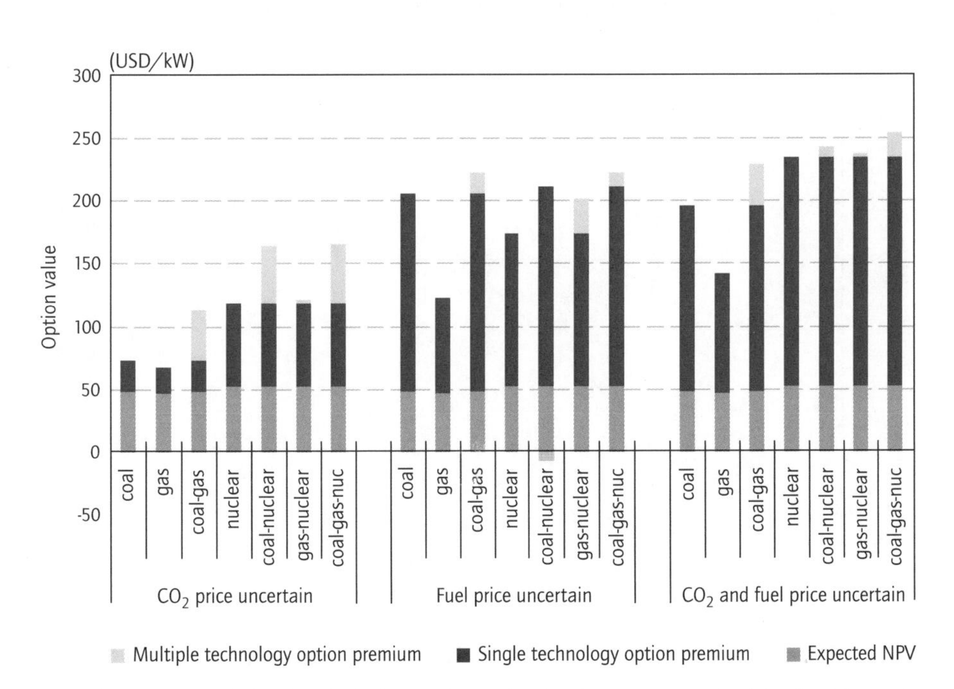

The option values are calculated on the basis of a mixture of coal and gas on the margin of the electricity supply merit order, and on both carbon price and fuel price uncertainties.

more technologies together. The red and beige sections taken together are equivalent to the difference between the red curve and blue line in Figure 31.

It is important to note when interpreting Figure 32 and Figure 33 that the option values shown here are not a direct measure of the investment threshold. The threshold required to stimulate investment would be higher than the values shown here, as can be seen from the results in Chapter 3.

When only CO_2 prices are uncertain, we see that the option value for coal and gas as individual investments is quite low, but that when the two technologies are considered together, there is an appreciable increase in option value. This technology interaction effect follows the same pattern as was described in the first example in Chapter 3. It arises because the revenues from coal and gas plants move in opposite directions to a change in CO_2 price, so this increases the

FIGURE 33

Option values when coal is always on the margin

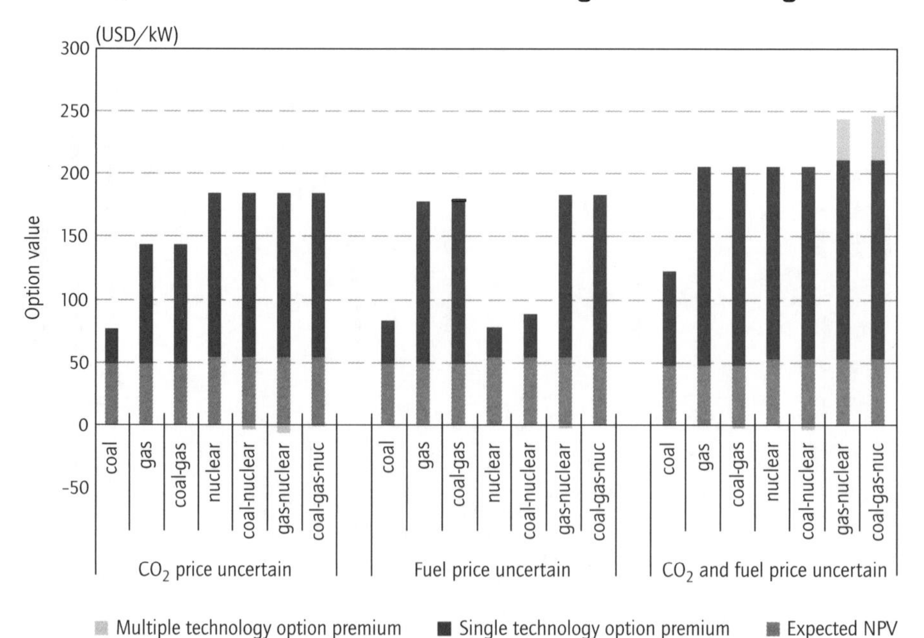

Option values when electricity prices are determined purely by coal plant (100% coal on the margin), taking into account carbon price and fuel price uncertainties.

option value of waiting. However, for the two-way coal and gas decision, fuel price uncertainty has a much stronger contribution to the overall option value than CO_2 price uncertainty, and this tends to dilute the technology interaction effect, as can be seen in the set of results on the right-hand side of Figure 32, which shows option values under combined CO_2 and fuel price uncertainty.

The addition of nuclear into the choice does not seem to add a great deal of additional option value. The additional technology interaction effects of the three-way coal, gas and nuclear decision are no stronger than the two-way decisions. This is because the two-way decisions already provide most of the optionality associated with having available investment options that move in opposite directions to a change in CO_2 price.

For the three-way coal-gas-nuclear investment decision, the overall option value tends to be dominated by the option value for nuclear, which has the highest individual option value. This is equivalent to saying that the risk premium for the combined decision is driven by the risk premium for nuclear.

As we saw in Chapter 3, when a coal plant is on the margin, CO_2 price risks are significantly higher for gas and nuclear plants, whereas the risks are relatively low for a coal plant. Gas price risks are also high for a gas plant when a coal plant is on the margin, since the price variations are not passed through to the price of electricity. The only technology interaction effect appears to occur in the choice between gas and nuclear, and is driven by their different responses to gas-price risks.

APPENDIX 3
COMPANIES CONSULTED DURING THE STUDY

We would like to acknowledge and thank the following the companies for their contributions of time and ideas to this study. Particular thanks also go to EPRI, Enel, E. ON UK, RWE npower and the governments of Canada, the Netherlands and the UK for their sponsorship of this project.

American Electric Power

Australian Gas and Light

Bank of Scotland

BP

Centrica

Climate Change Capital

Development Bank of Japan

EdF

Electric Power Research Institute

Enel

Eon-Energie

E. ON UK

Insight Investment

Origin Energy

RWE npower

Siemens

Southern Company

TransAlta

REFERENCES

Beaver, R. (1993). "Structural comparison of the models in EMF 12", *Energy Policy*, March 1993, 21(3): 238-248.

Birge, J. and Louveaux, F. (1997). *Introduction to Stochastic Programming*, Springer: New York.

Bunn, D. and Paschentis, S. (1986). "Development of a stochastic model for the economic dispatch of electric power", *European Journal of Operational Research*, 27, pp179-191.

Caldeira, K.; Jain, A.K. and Hoffert, M. (2003). "Climate sensitivity uncertainty and the need for energy without CO_2 emission", *Science*, Vol 299 28 March.

Dixit, A.K. and Pindyck, R.S. (1994). *Investment under Uncertainty*. Princeton University Press: New Jersey.

Department of Trade and Industry, United Kingdom (DTI) (2006). "The energy challenge". *Energy Review*, July 2006.

Energy Research Centre for the Netherlands (ECN) (2005). «CO_2 price dynamics: The implications of EU emissions trading for the price of electricity», ECN-C–05-081, The Netherlands.

Electric Power Research Institute (EPRI) (1999). "A framework for hedging the risk of greenhouse gas regulations", EPRI, Palo Alto, CA: TR-113642.

Epaulard, A. and Gallon, S. (2000). "La France doit-elle investir dans une nouvelle technologie nucléaire? La valorisation du Project Nucléaire EPR par la Méthode des Options Réelles", *Revue de l'Energie*, No. 515 pp 144-157.

Forest, C.E. *et al.* (2002). "Quantifying uncertainties in climate system properties with the use of recent climate observations", *Science*, 4 January, Vol 295 no 5552 p113-117.

Frame, D. *et al.* (2005). "Constraining climate forecasts: the role of prior assumptions", *Geophysical Research Letters*, Vol 32.

Helm, D. and Hepburn, C. (2005). "Carbon contracts and energy policy: an outline proposal", available at www.dieterhelm.co.uk/publications/CarbonContractsOct05.pdf.

Hepburn, C. *et al.* (2006). "Auctioning of EU ETS phase II allowances: how and why?" *Climate Policy* 6, p137-160.

International Energy Agency (IEA) (2002). "Beyond Kyoto", IEA/OECD: Paris.

IEA (2003). *World Energy Investment Outlook*, IEA/OECD: Paris.

IEA (2004a). *World Energy Outlook*, IEA/OECD: Paris.

IEA (2004b). *Prospects for Carbon Capture and Storage*, IEA/OECD: Paris.

IEA (2006a). "Light's labour's lost: Policies for energy-efficient lighting, in support of the G8 plan of action", IEA/OECD: Paris.

IEA (2006b). Energy *Technology Perspectives: Scenarios and Strategies to 2050*, IEA/OECD: Paris.

IEA (2006c). *World Energy Outlook*, IEA/OECD: Paris, (forthcoming/in press).

IEA (2006d). "Impact of climate change policy uncertainty in power investment", a working paper for the IEA website, LTO/2006/02, October, available at http://www.iea.org/Textbase/publications/free_new_Desc.asp?PUBS_ID=1824

IEA (2007). "Modeling investment risks and uncertainties with real options approach", a working paper for the IEA website, LTO/WP/2007/01, February, available at
http://www.iea.org/Textbase/publications/free_new_Desc.asp?PUBS_ID=1857

Ilex Energy Consulting (2004). «Impact of the EU ETS on European Electricity Prices», A report to the Department of Trade and Industry (DTI), Oxford, United Kingdom.

Intergovernmental Panel on Climate Change (IPCC) (2001). *Third Assessment Report: Climate change 2001: Synthesis report*, Cambridge University Press: UK.

IPCC (2005). *Carbon Dioxide Capture and Storage*, IPCC Special Report, available at www.ipcc.ch/activity/ccsspm.pdf.

Ishii, J. and Yan, J. (2004). "Investment under regulatory uncertainty: US electricity generation investment since 1996", Center for the Study of Energy Markets, University of California Energy Institute, Working Paper 127, March 2004.

Ismer, R. and Neuhoff, K. (2006). "Commitments through financial options", Electricity Policy Research Group working paper, available at
www.electricitypolicy.org.uk/pubs/wp/eprg0625.pdf.

Jacoby, H.D. (2004). "Informing climate policy given incommensurable benefits estimates", *Global Environmental Change*, 14, 287-297.

Kann, A. and Weyant, J.P. (2000). "Approaches for performing uncertainty analysis in large-scale energy/economic policy models", *Environmental Modeling and Assessment*, 5 pp29-46.

Kiriyama, E. and Suzuki, A. (2004). "Use of real options in nuclear power plant valuation in the presence of uncertainty with CO_2 emission credit", *Journal of Nuclear Science and Technology*, 41(7) p756-764.

Lambrecht, B. and Perraudin, W. (2003). "Real options and pre-emption under incomplete information", *Journal of Economic Dynamics and Control*, 27 619-643.

Louveaux, F. and Smeers, Y. (2005). "A stochastic model for electricity generation" in Haefele (ed.), *Proceedings 11 ASA/IFAC Symposium on Modelling on Large Scale Energy Systems*, Pergamon: Oxford.

Netherlands Bureau for Economic Policy Analysis (CPB) Memorandum (2003). «Emission Trading and the European Electricity Market», The Netherlands.

Nuclear Energy Agency/International Energy Agency (NEA/IEA) (2005). *Projected Costs of Generating Electricity*, NEA/IEA/OECD: Paris.

Nordhaus, W. D. and Popp, D. (1997). "What is the value of scientific knowledge? An application to global warming using the PRICE model", *The Energy Journal*, Vol 18, No. 1.

Papathanasiou, D. and Anderson, D. (2000). "Uncertainties in responding to climate change: On the economic value of technology policies for reducing costs and creating options", *Imperial College Centre for Energy Policy and Technology* (Draft May 1), available at www.iccept.ic.ac.uk/pdfs/techpol.pdf.

Pindyck, R. S. (1999). "The long-run evolution of energy prices", *The Energy Journal*, 20, 2; ABI/INFOM Global, pp1-27.

Philibert, C. (2006). "Certainty vs. ambition: Economic efficiency in mitigating climate change", IEA working paper series LTO/2006/03, available at www.iea.org/textbase/papers/2006/rb_certainty_ambition.pdf.

Pizer, W. (2002). "Combining price and quantity controls to mitigate global climate change", *Journal of Public Economics*, 85, pp409-434.

Reinaud, J. and Baron, R. (forthcoming). «Interaction between CO_2 and electricity prices - impacts on European industry's electricity purchasing strategies», Information paper, IEA/OECD: Paris.

Rothwell, G. (2006). "A real options approach to evaluating new nuclear power plants", *The Energy Journal*, Vol. 27 No. 1 p37.

Schneider, S.H.; Kuntz-Duriseti, D. and Azar, C. (2000). "Costing non-linearities, surprises and irreversible events", *Pacific and Asian Journal of Energy*, 10(1) 81-106.

Shuttleworth, G.; Gammons, S. and Hofer, P. (2005). "Methodology for Measuring CO_2 Pass-Through: A Report for EnergieNed", available at http://www.energiened.nl/Content/Home/HomePublic.aspx.

Standard and Poor's (2006). "Security of energy supply: a credit survey of Europe's energy infrastructure, supply availability, and exposure to gas imports", *October 2006*.

Stern, N. (2006). "The economics of climate change", Stern Review on the Economics of Climate Change, available at www.hm-treasury.gov.uk/independent_reviews/stern_review_economics_climate change.

Trigeorgis, L. (1996). Real Options: *Managerial flexibility and strategy in resource allocation*, MIT Press: Boston.

Trigeorgis, L. (1991). "Anticipated competitive entry and early pre-emption investment", *Journal of Economics and Business*, 43, 143-156.

Webster, M. (2002). "The curious role of 'learning' in climate policy: Should we wait for more data?", *The Energy Journal*, Vol 23, No. 2 p97.

Webster, M. et al. (2003). "Uncertainty analysis of climate change and policy response", *Climatic Change*, 61: 295-320.

White, A. (2005). "Concentrated power", *Public Utilities Fortnightly*, February 2005.

Yohe, G., Andronova, N. and Schlesinger, M. (2004). "To hedge or not against an uncertain climate future?", *Science*, Vol 306 15 October.

Yang M. and Blyth W. (2007). "Modeling Investment risks and uncertainties with Preal options approach" an IEA working paper. LTO/WP/2007/01 February.

IEA PUBLICATIONS, 9, rue de la Fédération, 75739 PARIS CEDEX 15
PRINTED IN FRANCE BY STEDI MEDIA
(61 2007 141 P1) ISBN13 : 978-92-64-03014-5 - 2007